Genetics and Sports
2nd, revised and extended edition

Medicine and Sport Science

Vol. 61

Series Editors

Dennis J. Caine Grand Forks, N.Dak.
Andrew P. Hills Brisbane, Qld.
Timothy Noakes Cape Town

Genetics and Sports

2nd, revised and extended edition

Volume Editors

Michael Posthumus Cape Town
Malcolm Collins Cape Town

2 figures, and 6 tables, 2016

KARGER Basel · Freiburg · Paris · London · New York · Chennai · New Delhi · Bangkok · Beijing · Shanghai · Tokyo · Kuala Lumpur · Singapore · Sydney

Medicine and Sport Science
Founded 1968 by E. Jokl, Lexington, Ky.
Honorary Series Editors: J. Borms, Brussels; M. Hebbelinck, Brussels

Michael Posthumus, PhD
Senior Research Officer
Division of Exercise Science and Sports Medicine
Department of Human Biology
University of Cape Town
Cape Town (South Africa)

Malcolm Collins, BSc (Hons), PhD, FECSS
Professor and HOD
Department of Human Biology
Faculty of Health Sciences
University of Cape Town
Cape Town (South Africa)

Library of Congress Cataloging-in-Publication Data

Names: Posthumus, Michael. | Collins, Malcolm, 1965- author.
Title: Genetics and sports / volume editors Michael Posthumus, Malcolm Collins.
Other titles: Medicine and sport science ; v. 61.
Description: 2nd, revised and extended edition. | Basel ; New York : Karger, 2016. | Series: Medicine and sport science ; Vol. 61 | Includes bibliographical references and indexes.
Identifiers: LCCN 2016017242| ISBN 9783318030105 (hard cover : alk. paper) | ISBN 9783318030112 (electronic version)
Subjects: | MESH: Sports | Genetic Phenomena | Athletic Performance--physiology | Exercise--physiology | Athletic Injuries--genetics | Genetic Techniques
Classification: LCC RC1235 | NLM QT 260 | DDC 599.93/5--dc23 LC record available at https://lccn.loc.gov/2016017242

Vol. 54 (1st edition)
Genetics and Sports
Editor: M. Collins, Cape Town, South Africa
VIII + 200 p., 11 fig., 12 tab., hard cover, 2009. ISBN 978-3-8055-9027-3

Bibliographic Indices. This publication is listed in bibliographic services, including Current Contents®.

www.karger.com
Printed in Germany on acid-free and non-aging paper (ISO 9706) by Kraft Druck GmbH, Ettlingen
ISSN 0254–5020
e-ISSN 1662–2812
ISBN 978–3–318–03010–5
e-ISBN 978–3–318–03011–2

Contents

Preface

The first edition of this book, *Genetics and Sport*, covered a number of topics in an attempt to obtain an integrated and holistic understanding of the field of sports and exercise genomics. Since the publication of the original edition in 2009, research in the field of genomics has continued to increase exponentially. Most notably, as the cost of whole-genome screening technology has become more affordable, we have seen a rise in the amount of genome-wide association studies. Other new technologies, such as copy number variants, next-generation sequencing technologies and epigenetic profiling, have further highlighted the complexities of trying to decipher exactly how the human genome regulates interindividual variation in athletic performance, response to training and susceptibility to injury as well as exercise-associated illnesses. Specific to the field of sports and exercise medicine, these technologies have only just started to contribute to our understanding of the interindividual differences in these traits (phenotypes). This book, the second edition in the series *Genetics and Sport*, provides an update on recent developments within the field.

The introductory chapter outlines the basic methodologies of genetic association studies and discusses the technological advances in genomics. Noteworthy advances include an increased appreciation for the potential role of the epigenome in performance and injury. This is followed by an update in the often misunderstood debate of nature (genetics) versus nurture (common environmental effects such as training, diet, etc.) in determining athletic ability. This chapter explains how the environment interacts with the genome and how the traditional argument of nature versus nurture has become irrelevant as our knowledge advances.

When discussing the genetics of athletic performance, it is important to consider that in addition to outcomes of performance, adaptation to training is also an extremely important phenotype which will translate to improved performance within an athlete. The following chapter reviews the recent advances in the genetics of exercise training adaptation. Despite a large amount of recent research, including further high-quality genome-wide approaches, such as the recent study produced from the HERITAGE cohort, the identification of genetic factors underlying the adaptations of various traits to exercise training remains unmet and elusive. Perhaps future mRNA transcriptome and epigenetic work may prove why discovering repeatable genetic factors has been so challenging. Conclusions within the chapter reviewing the advances in the genetics of performance are very similar. Several genetic markers (155 to be exact) have been identified to date; however, the association of the majority of these markers has not been replicated (approx. 80%) in independent samples, raising the possibility that many of these markers may be the product of false-positive findings. As the next chapter on the future of genomic research describes, a future effort to elucidate the genetic contribution to both performance and adaptation to training requires large, multicentre collaborative projects with sound experimental designs.

There is now mounting evidence that genetics plays an important role in both sport injury predisposition and other sports-related phenotypes. The evidence is reviewed within two subsequent chapters. Initially, the studies identifying genetic risk factors for musculoskeletal soft-tissue injuries are discussed. Since the genes associated with injury code for structural and/or regulatory components of musculoskeletal soft tissue, it is likely that these genetic variants affect the tissue mechanics. Thus, the subsequent chapter further reviews the genetics of other exercise-related phenotypes which also affect changes in tissue mechanical properties.

The final chapter discusses the practical and ethical considerations regarding genetic tests to predict performance and/or risk of exercise-related injury or illness. It is important that scientists and practitioners understand the limitations of the available genetic tests, as well as the ethical concerns and importance of counselling before and after genetic testing.

Michael Posthumus, Cape Town
Malcolm Collins, Cape Town

Posthumus M, Collins M (eds): Genetics and Sports, ed 2, revised, extended.
Med Sport Sci. Basel, Karger, 2016, vol 61, pp 1–14 (DOI: 10.1159/000445237)

Core Concepts in Human Genetics: Understanding the Complex Phenotype of Sport Performance and Susceptibility to Sport Injury

William T. Gibson

Department of Medical Genetics, Child and Family Research Institute, University of British Columbia, Vancouver, B.C., Canada

Abstract

High-throughput sequencing of multiple human exomes and genomes is rapidly identifying rare genetic variants that cause or contribute to disease. Microarray-based methodologies have also shed light onto the genes that contribute to common, non-disease human traits such as hair and eye colour. Sport scientists should keep in mind several things when interpreting the literature, and when designing their own genetic studies. First of all, most genetic association methods are more powerful for detecting disease phenotypes (such as susceptibility to injury) than they are for detecting healthy phenotypes (such as sport performance). This is because there are likely to be many more biological factors contributing to the latter, and the effect size of most of these biological factors is likely to be small. Second, implicating a particular gene in a human phenotype like athletic performance or injury susceptibility requires an unbiased population data set. Third, new types of non-coding biological variability continue to be uncovered in the human genome (e.g. epigenetic modifications, microRNAs, etc.). These other types of variability may contribute significantly to differences in athletic performance.

Athletic performance has long been known to be modifiable by dedicated training [1]. However, some individuals appear to be naturally gifted with athletic ability. Their performance is above average, even prior to training, and their performance after training is consistently excellent. The degree to which athletic potential is predetermined by inherited traits, and the degree to which the response to training can be predicted prior to the training itself taking place, has excited much debate – debate that is often framed as 'nature versus nurture'. However, there is very likely to be a strong interaction between the genetic ('nature') and the environmental ('nurture' – diet, exercise and other training) contributions to sport performance. Studies that measure variation at the DNA level have suggested that certain specific genes are involved in athletic performance [for a review, see 2], though many association signals disappear when more stringent genome-wide significance is insisted upon [3]. We do not yet have a full picture of the spectrum of common polymorphisms and rare DNA variants that might affect athletic parameters like strength, endurance, coordination and recovery time. The relative contribution of common DNA variants and of rare DNA variants to a human trait is termed the 'genetic architecture' of the trait (table 1) [4]. Though many rare genetic diseases have a single gene as their major determinant, complex non-disease phenotypes are most likely influenced by multiple different types of DNA variants, both rare and common. This introductory review will aim to equip the reader with the necessary vocabulary to understand and interpret genetic studies targeted to sport performance and sport-related injury.

Key Genetic Concepts and Terms

The word 'phenotype' is used to describe a person's physical characteristics, either in broad terms (e.g. the ectomorph/mesomorph/endomorph somatotypes of so many decades ago) [5] or in more specific terms (e.g. an ultrarapid metabolizer of codeine) [6]. The word 'genotype' is used to describe a person's genetic characteristics. As with the phenotype, the genotype can be defined in broad terms (e.g. by continental ancestry) or in more specific terms (e.g. an R577R/R577X heterozygote for a common DNA variant in the *ACTN3* gene). Published evidence that links a particular phenotype to a particular genotype varies in quality: many genotype-phenotype correlations are published once and never replicated, because replication and real proof often require highly detailed statistical methods applied to a replication cohort with tens of thousands to hundreds of thousands of participants. Furthermore, the utility of the association data depends strongly on how specifically the phenotype can be defined. For example, dual-energy X-ray absorptiometry would provide higher-resolution information on the phenotype of muscle mass than would skinfold thickness measurements, which in turn would be preferable to body mass index alone. A robust correlation between physical traits and the genetic variation that underlies them is best achieved with high-resolution phenotyping paired with high-resolution genotyping.

Just as methods for phenotyping an athlete can vary in complexity and in resolution, methods for genotyping can also vary in quality and level of detail. One widely quoted statistic categorizes all humans as being 99.9% genetically identical to each other. However, in a genome of 6.9–7 billion base pairs, this would still leave 7 million base pairs of genetic difference between two randomly selected people. Given the

Table 1. Key terms in human genetics

Allele: An observed DNA sequence at a specific site in the genome, whether rare or common
Candidate gene association study (CGAS): A popular type of case-control study that relies on a prior hypothesis that variation within a particular gene influences the phenotype of interest; though inexpensive and intuitively appealing, CGASs have fallen out of favour because of their very high false discovery rate (poor replicability)
Copy number variant (CNV): A deletion or duplication of one or more genes such that a person has 0, 1, 3, 4 or more copies of a gene, whereas 2 copies are seen in the average individual
De novo mutation: A new mutation – one that has been proven through genetic studies to be present in a person's DNA yet is absent in both of his or her biological parents' genomes; though not inherited from either parent, a de novo mutation is still heritable – it can be passed on to children like any other genetic variant; not all de novo mutations affect the phenotype
DNA marker: A common variant that is physically linked to another variant of interest on a haplotype, and/or associated statistically to a physical trait, thereby 'marking' that variant or trait
Epigenome: Genomic modifications that affect its transcriptional accessibility without changing the DNA sequence; epigenetic modifications include methylation and wrapping of DNA around core histone proteins
Exome: The portion of the genome that is expressed as proteins: about 20,000 genes that occupy ~1% of the genome
Genome: The total DNA of an organism; the Genome Reference Consortium Human Build 38 lists 3,547,121,844 base pairs in the 'average' set of human chromosomes such that females are expected to have ~7 billion base pairs in their genome (which contains 2 sets of chromosomes); males are expected to have ~6.9 billion base pairs (a slightly smaller genome because the human Y chromosome is roughly 100 million base pairs smaller than the X) [4]
Genotype: The specific allele(s) observed in an identified individual
Genetic architecture: The combination of rare and common variants that contribute to a particular phenotype when added together; many rare diseases are only caused by rare mutations in one particular gene, though for some rare diseases different mutations in different genes may cause the same disease; the genetic architecture of common, complex diseases (e.g. heart disease, diabetes, dementia, cancer) typically includes multiple rare and common variants; the genetic architecture of sport performance is thought to be similarly complex
Genome-wide association study (GWAS): The major method now used to associate common human phenotypes with common genotypes; typically, GWASs use microarrays to interrogate millions of pre-identified single-nucleotide polymorphisms distributed across the genome of each study participant; sample sizes of 10,000–200,000 unrelated people are usually required
Haplotype: A series of genetic variants that exist all in a row on a specific chromosome
Indel: A genetic variant where one or more nucleotides can be present or absent at a genomic address (e.g. CCACC vs. CCCC at the same location); indels may be common or rare at a particular site
Locus: An identifiable genetic address within a particular genome
Microarray: A type of whole-genome analysis that genotypes millions of genomic addresses without sequencing between them; microarrays can detect many single-nucleotide polymorphisms and can detect large duplications and deletions of DNA; they typically do not detect rare variants, mutations or small indels (unless designed specifically to do so)
Mitochondrial genome (mtDNA): The small but powerful DNA inside the cell's energy-producing organelles; there are 16,569 base pairs in the mtDNA and 13 coding genes; most mitochondrial proteins are coded by the nuclear genome and imported into the mitochondria, so many mitochondrial phenotypes exhibit Mendelian inheritance and not pure matrilineal inheritance

Table 1. Continued

Pathogenic mutation: A genetic variant that has been causally associated with disease; dominant pathogenic mutations cause disease when present on one chromosome only, whereas recessive pathogenic mutations only cause disease when present on both chromosomes
Phenotype: The physical feature(s) observed or measured in an identified individual; phenotypes may include the presence or severity of disease, or of quantifiable non-disease traits such as muscle strength, maximal running speed, endurance, etc.
Polymorphism: A DNA variant that is common in a population (typically observed in more than 1% of individuals)
Rare variant: A DNA variant that is uncommon in a population (typically observed in fewer than 1% of individuals)
Single-nucleotide polymorphism (SNP): A single-nucleotide polymorphism that is common in a population, wherein different nucleotides can be seen at a particular genomic address (e.g. C vs. A)
Transcriptome: The sum total of the ~200,000 RNA transcripts from all transcribed genes in the genome (including non-coding RNAs); more often used in a tissue-specific way to describe the subset of these RNA transcripts identified in a particular sample (e.g. skeletal muscle, cardiac muscle)
Whole exome sequencing (WES): A high-resolution technique that sequences the entire coding DNA of an individual (~20,000 genes)
Whole genome sequencing (WGS): A high-resolution technique that sequences the entire DNA of an individual; though relatively expensive, it is expected to detect most rare variants and all common single-nucleotide polymorphisms

observed diversity of phenotype between humans, even a 0.1% difference between genomes must be highly significant. The 99.9% figure is given here as a generic example because many readers will have heard or read about it. In fact, both the accuracy and the significance of this 99.9% identity statistic have been disputed by experts in genomics for many years. Real genetic differences between two randomly selected healthy humans would include not only substitutions of one letter for another in the genetic code (single-nucleotide polymorphisms, SNPs, and rare variants), but also multiple insertions and deletions (small 'indels' as well as larger ones) as well as other complex rearrangements of the DNA base pairs. No single number properly captures this variability. Interested readers may wish to compare the paper that describes the initial draft of the human genome [7], with those that describe the completion of the 1,000 Genomes Project and similar consortia [8–10].

Most genes are physically located in the nuclear compartment of the cell, though the energy-processing mitochondria have their own mini-genome (mtDNA) that has several genes of its own. The term 'gene' is used here to refer to a discrete sequence of DNA that produces a product. This could be a messenger RNA and protein, a non-coding microRNA, a tRNA, or any other biologically active molecule. If different sequences of the same gene coexist in a population, that gene is said to have different forms termed 'alleles'. An allele is a variant of an identified gene – a situation where more than one specific sequence is possible for a particular gene. People who have different alleles for a particular gene are said to be heterozygous, and those who have identical alleles on both chromosomes are said to be homozygous. The genome is the sum total of all nuclear genes, intervening DNA sequences within and between genes, and mitochondrial sequences present in an individual. The term itself can be used to refer to one unique series of genes (haploid genome), to the redundant series of genes present in an organism (diploid genome) or to the collection of genes present in a group of individuals under study (population or species genome).

A string of 4 chemical nucleotide bases, referred to by the letters A, C, G and T, makes up a DNA sequence. By default, the most common combination of letters (such as GCGTTA) found in a particular region of the genome is taken to be the normal sequence (also known as the 'canonical' or the 'reference' sequence). Differences from that order of nucleotides are considered to be DNA variants, and many variants are possible. In the example above, the sequence GCGATA would be considered a variant, because it differs at the third position from the right: an A exists where a T would normally be found. Variants that change one chemical letter are referred to as rare variants or polymorphisms (from the Greek for 'many forms'), depending on their frequency in the population. If the GCGTTA sequence were present in 55% of the sequences in a study population, and the GCGATA sequence were present in 45%, the 'T' variant would be considered the 'normal variant', the 'A' variant would be considered to be the minor allele, and the position in the genome where these alleles were found would be considered to harbour a T/A 'SNP' at that site. If, on the other hand, 99.95% of human sequences at that site were found to be GCGTTA and 0.05% were found to be GCGATA, the minor-frequency 'A' allele would be considered a rare variant and possibly termed a 'mutation'. Debate exists whether the term 'mutation' should be reserved exclusively for rare DNA variants that have major adverse effects on physical characteristics (pathogenic mutations) or perhaps for new DNA variants that have changed for the very first time in a specific generation (de novo mutations). For rare diseases wherein a DNA sequence variant is 100% associated with disease, the use of 'mutation' is well accepted. In population studies, the term 'mutation' is often applied to any rare DNA variant(s) that are seen at a frequency of 1% or less, whether or not they affect the phenotype. In this review, no attempt will be made to settle this debate. The terms 'rare variant' and 'mutation' will both be used, because both are found frequently in the literature.

A haplotype or haploid genotype refers to a known series of DNA variants that occur on the same chromosome, and thus on the same strand of DNA. Haplotypes are used often in genetic studies; they may contain a combination of SNPs, rare variants and/or other DNA sequences such as insertions of base pairs, deletions that remove some nucleotides, inversions where the DNA letters appear in reverse order, and other complex rearrangements that include a scrambling of the normal order of DNA letters.

The existence of haplotypes allows researchers to make a sort of educated guess about a person's genotype at a variety of different positions – if a group of SNPs and/

or rare variants are seen very frequently together on the same chromosome, then genotyping a specific subset of these variants, sometimes even a single variant, can provide sufficient information to infer the presence of the other variants. Full sequencing is not always necessary if variants that 'tag' the appropriate haplotype can be genotyped as proxies for a region of interest. The advent of genome-wide analysis using microarrays bearing a large number of tagging variants has enabled newer genome-wide association studies (GWASs) that have replaced older methods such as twin studies and candidate gene association studies (CGASs). Because haplotypes tend to be passed down stably from one generation to another, they can also be used as markers of ancestral origin, which allows important corrections for ancestry to be made in a GWAS that cannot be done with traditional CGAS methods.

Because DNA (and its attendant genes, variants and haplotypes) is passed down from one generation to another, observed physical characteristics that are passed down from one generation to another are, for the most part, believed to have a genetic origin. When a physical trait or phenotype is described as being 'Mendelian', then it has been observed to follow Gregor Mendel's laws of inheritance (dominant, recessive or X-linked). Genetic diseases usually follow Mendel's laws, but it is important to recognize that most physical traits associated with athletic performance will *not* be Mendelian traits. Mendelian traits are transmitted with high frequency through families, which means that the risk to an immediate relative (parent, sibling or offspring) is rather high, in the order of 25–50%. Dominant traits are passed directly from parent to child; the causative allele is inherited from one parent who also has the physical trait (whether it be a healthy or a disease-related trait). Siblings and offspring of someone with a dominant trait have a 50% risk of inheriting the same allele and hence of manifesting the same trait. Recessive traits appear in the children of two carrier parents; one mutant allele must be inherited from each (unaffected) parent, leading to a 25% risk to siblings. A special exception exists for sex-linked or X-linked recessive traits; these show up in 50% of male siblings, with a 25% risk to all siblings when males and females (themselves at 50% risk for inheriting the gene but low risk for manifesting the phenotype) are considered together. For Mendelian traits, the physical trait correlates very highly with the susceptibility allele. Often, 100% of individuals who inherit the mutant gene(s) will manifest the trait.

Redundant though it may be to state herein, complex traits are not so simple. Complex traits are those physical characteristics that are influenced by both genetic and environmental factors such as diet, intrauterine environment and activity level. Complex traits are often quantitative traits, and in the normal population these terms are largely equivalent. Quantitative traits are physical characteristics for which a range of measurable values are possible. Quantitative traits include height, weight, forced vital capacity, cardiac stroke volume, etc. By contrast, qualitative traits are discrete physical characteristics that are either present or absent. Most complex traits encountered in sport science are assumed to be quantitative traits.

Genetic mapping is the process of assigning biological traits to genes and/or of assigning genes to specific regions of chromosomes. Chromosomes are packages of DNA that are visible during cell division. Because chromosomes are easily studied, they have become a convenient handle for gene mapping. Genes are mapped onto a chromosomal space much like a postal address might be mapped onto a physical space. A distinction should be made here between genes (defined above) and DNA markers. DNA markers (also known as genetic markers or sequence tags) are used as reference points to describe the location of genes relative to one another. Just as a landmark on a map can be any feature of significance (a bridge, hill, church, pub, intersection, etc.), a marker is merely a DNA sequence whose address is known: it need not have an identifiable RNA or protein product. SNPs are commonly used as DNA markers; millions of different SNPs have been identified at millions of different DNA addresses. SNPs are usually assigned an 'rs' (reference SNP) number, such as rs1815739. It is easy to test whether an individual carries a particular SNP, and to determine how many copies of that SNP he or she has. For the above reasons, SNPs are well suited for genetic mapping studies that associate physical characteristics (phenotype) with different genotypes. Most SNPs do not have a known biological function, though some SNPs in the coding portion of DNA do affect proteins. The rs1815739 mentioned above changes a C to a T in the DNA, which converts a 3-letter codon for arginine into a stop codon, terminating the protein chain of α_3-actinin at position 577 of the protein. This type of variant is called a 'nonsense mutation' because it ends the 'sense' of the protein chain at that point. Nonsense mutations are often contrasted to 'missense' mutations that change a single amino acid link in the chain to a different amino acid – one that often puts a differently shaped link into the chain and affects the chain's folding. Other variants such as indels can shift the way the DNA sequence is read, ending the normal chain of amino acids and replacing the remainder with a different (often shorter) chain of amino acids.

A locus is a region of the genome that has a physical trait, DNA marker, allele or gene associated with it. A locus may be a theoretical entity and is not necessarily a gene, though a genetic locus is often assumed to contain a gene that somehow works on the physical trait under study. The locus need not code for a protein or an RNA – it could be a non-coding sequence that controls the activity of one or more genes (e.g. by binding an activator protein). The locus can also be a 'neighbour' to a nearby sequence that is the actual determinant of the trait. If epidemiological data associate a physical trait such as height with a particular chromosome, that region is said to be a 'locus' for height [11]. It may be helpful to think of a locus as a low-resolution hypothesis: once data are sufficient to associate a physical trait with a region of the genome, a locus is defined and the relevant gene(s) or other sequences in the region are sought using higher-precision methods. A locus is a type of genetic 'neighbourhood', used as a term of reference until the actual address is found in the DNA sequence. Additional information, concepts and excellent diagrams may be found in the review by Attia et al. [12].

Genome-Wide and Candidate Gene Association Studies

The recent sequencing of the human genome has generated an extraordinary number of tools for DNA analysis, including several million SNPs [7]. Notable successes in the identification of rare Mendelian single-gene disease traits have spurred widespread enthusiasm for DNA analysis as a predictive factor for disease risk. The rapid drop in costs for SNP genotyping and DNA sequencing has spurred an equally rapid growth in studies that attempt to correlate DNA sequence variants with variation in measurable physical characteristics (e.g. height [11], body mass index [13] and $VO_{2\,max}$ [14]). Though the usefulness of DNA sequence analysis for finding mutations that cause rare genetic disorders has been proven, as has the usefulness of genome-wide SNP profiling for many complex traits, the relative contribution of genetics to sport-related phenotypes remains to be clarified and is likely to vary with the sport phenotype under study.

An association study is an epidemiological tool that measures the degree to which an identifiable factor explains the risk for a particular disease. This may be an environmental factor (e.g. cigarette smoking) or a genetic factor (e.g. the rs1815739 SNP). Genetic traits can also be associated with 'risk' for improved health such as high-performance athleticism. Typically, association studies either employ a candidate gene approach or a genome-wide approach. Those that preselect a specific gene or genes and look at how frequently common or rare variants in these genes co-occur with a physical trait or phenotype are termed CGAS. The genes in question are chosen as candidates based on prior knowledge of what they are already thought to do, on the assumption that that specific function is both relevant to the disease under study and also non-redundant (i.e. not compensated fully by another gene or genes elsewhere in the genome). Studies that examine DNA markers at many positions across all chromosomes are termed GWAS. GWASs employ thousands to millions of SNPs, each of which is 'typed' (short for 'genotyped') in a large population of individuals. Importantly, study participants must also have their physical characteristics measured directly (e.g. height, body mass index, $VO_{2\,max}$, as above). GWASs are much more expensive than CGASs, in part because of the cost of phenotyping and also because of the specialized statistical expertise that is required to assure quality data and to adjust for or eliminate confounding variables.

The ultimate goal of both the CGAS and GWAS is to measure the strength of genotype-phenotype correlations, which may be strong, weak or absent. Rare genetic diseases typically show extremely strong genotype-phenotype correlations: a child who tests positive for 2 copies of the p.Phe508del mutation in the *CFTR* gene is extremely likely to develop cystic fibrosis later in life [15], and specific mutations in human myostatin correlate strongly with increased muscle mass [16]. However, it is important to know that most genotype-phenotype correlations for non-disease traits are rather weak. Red hair and green eyes are often seen in individuals who carry polymorphisms at the *MC1R* gene, but DNA variation at this locus does not guarantee that someone will develop these traits, nor is *MC1R* the only gene that controls hair and eye pigmentation [17].

Because fewer markers must be genotyped in CGASs than in GWASs, the former are less expensive and are often done as a first step to assess the risk of sport-related injury that might be imparted by a particular DNA variant [18]. An SNP in one of the collagen genes could be genotyped among swimmers with rotator cuff injuries and among swimmers without such injuries, to measure how strongly carrier status for that SNP explains the risk of injury. Equally well, a CGAS could be performed to assess the 'risk' for high performance [19]. An SNP in or near the gene for α-actinin 3 might be genotyped in order to obtain the frequency of this SNP among high-performance sprinters (the 'cases') compared to members of the general population or to endurance runners (the 'controls'). At face value such studies are attractive because of their low cost, but sport scientists should interpret them with caution. Though they are inexpensive and relatively straightforward to do, the statistical associations that they discover are usually not found to be biologically true when studied in more detail. There are a number of reasons for this high false discovery rate (reviewed elsewhere [20]): it is extremely hard to control for or eliminate all possible sources of bias in a CGAS. A major source of bias in many studies is the population genetic substructure: the historical mixing between human populations that differed in the prevalence of genetic and physical characteristics. Suffice to say that the actual ancestry of most individuals is much more complex than can be identified through questioning by a researcher, and this fact can lead to spurious statistical associations between ancient genetic markers and contemporary physical traits [21]. Other issues such as potential genotyping errors, blinded testing and reporting bias are also worth considering but are not dealt with in detail here.

Definitive proof that a candidate gene (or a variant in such a gene) really predicts risk for the phenotype of sports injury would require a long-term prospective study. DNA samples would have to be collected prospectively in a large cohort of athletes, who would then have to be followed over time until a sufficient number of them had developed the injury. Since this type of study is tremendously time-consuming and expensive, an intermediate standard of proof is typically sought, wherein the findings of a preliminary association study are replicated in a larger retrospective population. For example, an association between SNPs in collagen genes and rotator cuff injuries among 100 cases and 250 controls from one academic centre in a large city might be considered to be 'replicated' if, on review of a national cohort, carriers of the at-risk variant were found to have a higher prevalence of rotator cuff injuries in the previous 10 years than were non-carriers. Typically, however, when replication studies like this one are done, the initial statistical association is no longer verifiable. Part of the difficulty likely lies in the fact that it is easy to come up with a biologically plausible hypothesis about how a particular DNA variant *might* influence disease. Whether the variants that *actually exist* in the human population *truly do* influence disease is another matter entirely. The answer is unknown at the time the study is done, but the accumulated experience from a number of CGASs for a wide variety of biological traits suggests that this method is prone to a high false discovery rate (i.e. the positive predictive value is low) [12, 20].

GWASs are thought to be more robust to replication than are candidate gene studies. The former are often touted as being 'agnostic' or 'hypothesis free', since no prior assumptions are made about the specific gene or genes involved in susceptibility to disease (or in 'susceptibility' to high-performance athleticism). However, this is a misnomer because GWASs are not, in fact, hypothesis free. The hypothesis they test is whether there is biological variation in the human genome that is common enough to confer a population-attributable risk that is sufficiently strong to be detectable. In simpler terms, GWASs test the hypothesis that there is, in fact, an association *somewhere in the genome* [22]. GWASs have achieved notable successes with biological traits for which genetic susceptibility is believed to be relatively high [11, 13]. However, moving beyond the original risk-tagging SNP down to the specific gene that confers risk is non-trivial and requires specific expertise in the relevant area of cellular physiology [23]. In this context, it is particularly important to remember that most genotype-phenotype correlations for non-disease traits are rather weak. A corollary to this is that, for most performance-related traits and for most sport-related injuries, claims of genetic association should be weighed carefully and replicated in large independent populations before being accepted as valid [24]. Because of the large number of SNPs genotyped, GWASs perform a number of comparisons that is several orders of magnitude greater than those of most epidemiological studies; thus, it is appropriate to seek much lower p values (in the order of 5×10^{-8} instead of 0.05) [25, 26]. In fact, a recent analysis of multiple athletic cohorts identified rs558129 at *GALNTL6* as the only promising SNP signal for endurance athleticism, and the p value did not reach a true genome-wide threshold [3].

Single-Nucleotide Polymorphisms versus Copy Number Variants: New Discoveries, New Technologies and New Challenges

Though popular, GWASs do not actually cover all of the variability in the human genome. It has recently been discovered that a large amount of genetic difference between individuals is due neither to mutations nor to SNPs, but to duplications and deletions of genes or groups of genes [27, 28]. The groups of extra and/or missing DNA sequences on human chromosomes have been termed copy number variants (CNVs) [29], and they reveal a previously uncharacterized architecture of the genome. Thus, the size of one person's genome may differ substantially from that of other normal people, and from the theoretical reference human genome. The regions that differ may contain whole genes or entire groups of genes, such that people may vary not only in the sequence at specific detectable SNPs, but also in the number of genes that they have. It may be helpful to think of the presence of CNVs as representing a certain amount of 'wobble' in the size of the human genome.

The global significance of CNVs for normal physiology is not clear, and most CNVs are transmitted stably through families and cause no problems. However, CNVs are thought to hold significant potential as a source of phenotypic variation. Indeed, the impact of CNVs on human physical characteristics may well be greater than the impact of SNPs and of rare mutations, either at the individual or population levels. This is a new and exciting area of research; whether CNVs impact athletic performance or susceptibility to sport-related injury remains to be evaluated. Each person is estimated to have approximately 160 of these CNVs (80 per haploid genome), and more than 18 million base pairs per person are affected by insertions, deletions and other structural variants [8, 30]. The impact of a particular insertion or deletion CNV on physical characteristics depends on the gene(s) involved, which correlates with the size of the CNV and the gene density of the region where it occurs. At the time of this writing, almost 500,000 human CNVs have been described [31]. A relevant feature of these polymorphisms is that CNVs tend to be found outside functioning genes rather than within them. The fact that gene-rich areas are relatively depleted of CNVs suggests that CNVs occurring in or near genes tend to have a mildly harmful effect on those genes. If CNVs had no such effect, they would likely be distributed more randomly throughout the genome. Therefore CNVs that are found in, near or including genes should be more likely to have a biological effect than would a nearby SNP.

The Implications of Human Genomic Structure for Sport Science

Sport science researchers who are planning studies to elucidate the complex genetic architecture of sport phenotypes such as high performance or injury susceptibility should be aware of the following properties of the human genome. The proportion of the genome that is made up of functioning genes is believed to be less than 3% [32], and the average coding length of a human gene is approximately 1,500 bp or 1.5 kb [33]. Estimates of the frequency and distribution of SNPs in the human genome vary from between 1 in every 180th nucleotide pair to 1 in every 280th nucleotide pair, on average [34, 35]. Thus, it is likely that polymorphic SNPs will be present in or near any candidate gene(s) selected for study. Each person also has between 18 and 20 million base pairs affected by structural variants [8, 30], an estimate that is likely to rise as better methods are applied to their detection. Thus, study designs should not focus exclusively on SNPs in plausible candidate genes – they should incorporate methods that assess copy number variation of the gene of interest. Similarly, GWAS should incorporate assessment of CNVs throughout the whole genome of the study participants. This is becoming easier as software methods are developed to derive CNV data from microarrays that profile SNPs. Several microarrays are available on the market [36]; these arrays have different levels of resolution, which translates into the ability to detect insertions and deletions of different sizes. Detection accuracy may also be improved by using multiple independent computational approaches to analyse the data [37].

In addition to corroborating data from in vitro functional studies (such as expression levels for each identified allele), validation of any association findings will require replication in larger cohorts, ideally in one or more independent populations, and preferably with a prospective study design. Such stringent criteria are difficult to achieve, particularly among performance athletes where large sample sizes may not be available. At the very least, large control cohorts should be sampled from the same ethnocultural population as the athletes themselves, such that adequately narrow confidence intervals are available for the 'background' frequencies of the DNA variants studied. Population substructure may dramatically skew the results of CGASs [21], so markers of population ancestry should be genotyped where possible, particularly in populations where admixture is known to have occurred historically [38].

One major strength that sport science may draw upon is the fact that numerous tools are available to characterize the phenotype of performance athletes in detail, and to define the precise nature of the pathology in injured athletes. Detailed phenotyping, properly applied, may improve the power to detect a biological effect in a small cohort (relative to a large cohort that is studied in less detail). Similarly, detailed genotyping that incorporates DNA markers of population ancestry and assessment of structural variants should be the goal of studies designed to correlate performance traits with human genes.

Novel sequencing technologies on the horizon for sport science include whole-exome sequencing, whole-genome sequencing, transcriptome analysis and epigenetic profiling. The exome is the portion of the genome that codes for proteins. It is also the portion for which software algorithms can most easily estimate the possible effects of DNA variants on a protein's function, because they can use the genetic code to predict the changes to the protein's amino acid chain that are likely to result from a change in the coding sequence. Though whole-genome sequencing is available, prediction of the effects of non-coding variants on gene expression remains difficult, so most research teams prefer to use microarrays and/or exomes on a larger sample of participants, rather than to allocate their resources to whole genomes. Assessment of the transcriptome offers another approach, whereby the total amount of RNA for a large number of genes is quantified at once. Transcriptomic methods produce complex and multidimensional data that profiles a snapshot of which genes were being transcribed at the time the sample was collected. However, a major challenge faced by these methods is the fact that they require tissue biopsies, ideally more than one in order to compare the transcriptome between different physiologically relevant states (e.g. skeletal muscle prior to training, before exercise and after exercise). For this reason much transcriptomics work has focused on animal models [39]. The epigenome is postulated to play an important role in athleticism and injury repair as well; a cell's epigenome affects the transcriptional accessibility of its genome without alterations to the nucleotide sequence itself [40]. These modifications 'beside' the sequence include methylation of DNA and wrapping of the DNA strands around core histone proteins: they render the genome more compact and less active, or more open and more active.

In theory, epigenetic modifications could plausibly affect a particular metabolic pathway as much or more than a person's complement of rare variants, structural variants and SNPs [41]. However, the analysis necessary to pull a biologically relevant signal out of the data complexity offered by transcriptomes and epigenomes likely means that the required sample sizes will rival or exceed those of a GWAS.

Much remains to be learned about the key genetic (and also transcriptional and epigenetic) factors that influence common quantitative traits like sport performance, susceptibility to injury and recovery from injury. Methods developed to find gene-disease associations are increasingly being adapted to seek markers for non-disease traits; lessons learned from these gene-disease associations include the fact that new technologies must be applied to large, well-phenotyped cohorts in order to generate robust results [42].

References

1 Lortie G, Bouchard C, Leblanc C, Tremblay A, et al: Familial similarity in aerobic power. Hum Biol 1982; 54:801–812.
2 Ahmetov II, Fedotovskaya ON: Current progress in sports genomics. Adv Clin Chem 2015;70:247–314.
3 Rankinen T, Fuku N, Wolfarth B, Wang G, et al: No evidence of a common DNA variant profile specific to world class endurance athletes. PLoS One 2016; 11:e0147330.
4 Genome Reference Consortium Human Build 38, October 15 update. http://www.ensembl.org/index.html (accessed March 4, 2016).
5 Sheldon WH: The Varieties of Human Physique: An Introduction to Constitutional Psychology. New York, Harper, 1940.
6 Madadi P, Ross CJ, Hayden MR, Carleton BC, et al: Pharmacogenetics of neonatal opioid toxicity following maternal use of codeine during breastfeeding: a case-control study. Clin Pharmacol Ther 2009; 85:31–35.
7 Lander ES, Linton LM, Birren B, Nusbaum C, et al: Initial sequencing and analysis of the human genome. Nature 2001;409:860–921.
8 1,000 Genomes Project Consortium; Auton A, Brooks LD, Durbin RM, et al: A global reference for human genetic variation. Nature 2015;526:68–74.
9 UK10K Consortium; Walter K, Min JL, Huang J, et al: The UK10K project identifies rare variants in health and disease. Nature 2015;526:82–90.
10 Exome Aggregation Consortium, et al: Analysis of protein-coding genetic variation in 60,706 humans. http://biorxiv.org/content/early/2015/10/30/030338.
11 Wood AR, Esko T, Yang J, Vedantam S, et al: Defining the role of common variation in the genomic and biological architecture of adult human height. Nat Genet 2014;46:1173–1186.
12 Attia J, Ioannidis JP, Thakkinstian A, McEvoy M, et al: How to use an article about genetic association. A. Background concepts. JAMA 2009;301:74–81.
13 Locke AE, Kahali B, Berndt SI, Justice AE, et al: Genetic studies of body mass index yield new insights for obesity biology. Nature 2015;518:197–206.
14 Ahmetov I, Kulemin N, Popov D, Naumov V, et al: Genome-wide association study identifies three novel genetic markers associated with elite endurance performance. Biol Sport 2015;32:3–9.
15 Rowntree RK, Harris A: The phenotypic consequences of *CFTR* mutations. Ann Hum Genet 2003; 67:471–485.
16 Schuelke M, Wagner KR, Stolz LE, Hubner C, et al: Myostatin mutation associated with gross muscle hypertrophy in a child. N Engl J Med 2004;350: 2682–2688.
17 Sulem P, Gudbjartsson DF, Stacey SN, Helgason A, et al: Genetic determinants of hair, eye and skin pigmentation in Europeans. Nat Genet 2007;39:1443–1452.
18 September AV, Schwellnus MP, Collins M: Tendon and ligament injuries: the genetic component. Br J Sports Med 2007;41:241–246.
19 Bray MS, Hagberg JM, Perusse L, Rankinen T, et al: The human gene map for performance and health-related fitness phenotypes: the 2006–2007 update. Med Sci Sports Exerc 2009;41:35–73.
20 Attia J, Ioannidis JP, Thakkinstian A, McEvoy M, et al: How to use an article about genetic association. B. Are the results of the study valid? JAMA 2009;301: 191–197.
21 Marchini J, Cardon LR, Phillips MS, Donnelly P: The effects of human population structure on large genetic association studies. Nat Genet 2004;36:512–517.

22 Hoggart CJ, Clark TG, De Iorio M, Whittaker JC, et al: Genome-wide significance for dense SNP and resequencing data. Genet Epidemiol 2008;32:179–185.
23 Smemo S, Tena JJ, Kim KH, Gamazon ER, et al: Obesity-associated variants within FTO form long-range functional connections with IRX3. Nature 2014;507:371–375.
24 Chanock SJ, Manolio T, Boehnke M, Boerwinkle E, et al: Replicating genotype-phenotype associations. Nature 2007;447:655–660.
25 Ioannidis JP: Non-replication and inconsistency in the genome-wide association setting. Hum Hered 2007;64:203–213.
26 McCarthy MI, Abecasis GR, Cardon LR, Goldstein DB, et al: Genome-wide association studies for complex traits: consensus, uncertainty and challenges. Nat Rev Genet 2008;9:356–369.
27 Iafrate AJ, Feuk L, Rivera MN, Listewnik ML, et al: Detection of large-scale variation in the human genome. Nat Genet 2004;36:949–951.
28 Sebat J, Lakshmi B, Troge J, Alexander J, et al: Large-scale copy number polymorphism in the human genome. Science 2004;305:525–528.
29 Feuk L, Carson AR, Scherer SW: Structural variation in the human genome. Nat Rev Genet 2006;7:85–97.
30 Sudmant PH, Rausch T, Gardner EJ, Handsaker RE, et al: An integrated map of structural variation in 2,504 human genomes. Nature 2015;526:75–81.
31 Database of genomic variants. http://dgv.tcag.ca (accessed March 4, 2016).
32 Eyre-Walker A, Keightley PD: High genomic deleterious mutation rates in hominids. Nature 1999;397:344–347.
33 Crawford DC, Akey DT, Nickerson DA: The patterns of natural variation in human genes. Annu Rev Genomics Hum Genet 2005;6:287–312.
34 International Haplotype Consortium: A haplotype map of the human genome. Nature 2005;437:1299–1320.
35 Redon R, Ishikawa S, Fitch KR, Feuk L, et al: Global variation in copy number in the human genome. Nature 2006;444:444–454.
36 Carter NP: Methods and strategies for analyzing copy number variation using DNA microarrays. Nat Genet 2007;39:S16–S21.
37 Baross A, Delaney AD, Li HI, Nayar T, et al: Assessment of algorithms for high throughput detection of genomic copy number variation in oligonucleotide microarray data. BMC Bioinformatics 2007;8:368.
38 Tian C, Gregersen PK, Seldin MF: Accounting for ancestry: population substructure and genome-wide association studies. Hum Mol Genet 2008;17:R143–R150.
39 Capomaccio S, Vitulo N, Verini-Supplizi A, Barcaccia G, et al: RNA sequencing of the exercise transcriptome in equine athletes. PLoS One 2013;8:e83504.
40 Horsburgh S, Robson-Ansley P, Adams R, Smith C: Exercise and inflammation-related epigenetic modifications: focus on DNA methylation. Exerc Immunol Rev 2015;21:26–41.
41 Lindholm ME, Marabita F, Gomez-Cabrero D, Rundqvist H, et al: An integrative analysis reveals coordinated reprogramming of the epigenome and the transcriptome in human skeletal muscle after training. Epigenetics 2014;9:1557–1569.
42 Pitsiladis YP, Tanaka M, Eynon N, Bouchard C, et al: Athlome Project Consortium: a concerted effort to discover genomic and other 'omic' markers of athletic performance. Physiol Genomics 2016;48:183–190.

William T. Gibson, MD, PhD, FRCPC, FCCMG
Department of Medical Genetics, Child and Family Research Institute
University of British Columbia, 950 West 28th Avenue, Room A4-182
Vancouver, BC V5Z 4H4 (Canada)
E-Mail wgibson@cw.bc.ca

Posthumus M, Collins M (eds): Genetics and Sports, ed 2, revised, extended.
Med Sport Sci. Basel, Karger, 2016, vol 61, pp 15–28 (DOI: 10.1159/000445238)

Nature versus Nurture in Determining Athletic Ability

Xu Yan[a] · Ioannis Papadimitriou[a] · Ronnie Lidor[b] · Nir Eynon[a]

[a]Institute of Sport, Exercise and Active Living (ISEAL), Victoria University, Melbourne, Vic., Australia;
[b]Zinman College of Physical Education and Sport Sciences, Wingate Institute, Netanya, Israel

Abstract

This overview provides a general discussion of the roles of *nature* and *nurture* in determining human athletic ability. On the nature (genetics) side, a review is provided with emphasis on the historical research and on several areas which are likely to be important for future research, including next-generation sequencing technologies. In addition, a number of well-designed training studies that could possibly reveal the biological mechanism ('cause') behind the association between gene variants and athletic ability are discussed. On the nurture (environment) side, we discuss common environmental variables including deliberate practice, family support, and the birthplace effect, which may be important in becoming an elite athlete. Developmental effects are difficult to disassociate with genetic effects, because the early life environment may have long-lasting effects in adulthood. With this in mind, the fetal programming hypothesis is also briefly reviewed, as fetal programming provides an excellent example of how the environment interacts with genetics. We conclude that the traditional argument of nature versus nurture is no longer relevant, as it has been clearly established that both are important factors in the road to becoming an elite athlete. With the availability of the next-generation genetics (sequencing) techniques, it is hoped that future studies will reveal the relevant genes influencing performance, as well as the interaction between those genes and environmental (nurture) factors.

Nature versus Nurture or Nature Plus Nurture in Sports?

Human athletic ability is influenced by a number of factors, such as environmental, physiological, psychological, and sociocultural variables, and many others [1]. The term 'nature versus nurture', initiated many years ago, refers to whether heredity (nature) or the environment (nurture) has a greater impact on human development (behaviour, habits, intelligence, aggressive tendencies, athletic performance, etc.).

Possessing exceptional ability/abilities is, at least partially, attributed to one's genes. Talent is passed down from parents or grandparents to the next generation, and this can include intelligence or athletic performance. In fact, many athletes from the past and today are members of the same family.

However, there is also evidence that talent is learned and earned through extended and intense practice of a skill, namely 'No pain, no gain'. This idea is encapsulated in a golden rule made popular by the writer Malcolm Gladwell in his book [2]. This '10,000 h of practice' rule [3] is based on research by psychologist Anders Ericsson at Florida State University. The rule suggests that about 10,000 h of dedicated practice in your particular field is sufficient to bring out the best in you. In essence, Ericsson's theory suggests that sufficient practice in a particular skill can take anyone to a proficiency level equivalent to that of a top classical musician [3]. Ericsson and others do not deny that genetic limitations, such as those of height and body size, can constrain expert performance in areas like athletic performance. They argue, however, that there is not enough evidence to pinpoint the genes that directly influence elite athletic performance.

Studies on twins (involving both identical and non-identical twins) and family members, which implicate genetics as a significant factor pertaining to athletic performance and adaptation to exercise training (even after adjusting for environmental factors), have shown that 30–80% of the variation in various performance traits (phenotypes) can be attributed to genetics. An example is the well-known Health, Risk, Training and Genetic (HERITAGE) family study (which is broadly discussed in the next section), in which a number of exercise-related phenotypes are to be influenced by both environmental and genetic factors. Hence, the current paradigm is that athletic ability, in particular at the elite level, is influenced by components of both nature and nurture.

Genetics and Athletic Ability

Traditional Twin and Family Studies

The genetic basis of athletic ability has been studied for several decades. In the 1970s and 1980s, research into the effect of genetics on exercise and athletic performance was mainly conducted via twin and family studies. These studies demonstrated that genetics play a significant role in athletic performance and participation in exercise, even when adjusting for environmental effects [4]. More recently, a large-scale twin study composed of 37,051 twin pairs from 7 European countries suggested that the heritability of participating in leisure time exercise was between 48 and 71% [5]. A genome-wide linkage scan for athletic status reported a heritability of roughly 66% for athletic status in 700 British female dizygotic twin pairs [6]. In the HERITAGE family study, 742 healthy, sedentary participants, coming from families of white or black ancestry, underwent 20 weeks of standardized exercise training, and the heritability of several exercise-related traits was measured [7]. After adjustment for age, sex, body mass index, and baseline parameters, the heritability of the training responses for maximal oxygen uptake, submaximal exercise heart rate, and submaximal exercise capacity were estimated as 47% [8], 34% [9], and 26% [10], respectively. Muscle strength and mass have also been reported to be influenced by genetic factors. The estimated heritability for muscle strength and muscle mass, assessed by twin and family studies, varies from 31 to 78%, with large differences between muscle groups, contraction velocities, and muscle lengths [11].

The twin and family studies have been important landmarks in the field of sports genetics; however, studies using this approach did not identify individual genes that influence performance. Following the completion of the Human Genome Project in 2003, genetic research has evolved into more precise molecular DNA testing that is able to directly analyse the association between individual genes and athletic performance, and the hunt for sport gene/s was under way.

Identification of Specific Gene Variants Associated with Elite Athletic Ability

Genetic variants are present throughout the human genome and reflect differences in DNA sequence among individuals, groups, or populations. Genetic variants are common (they are the opposite of 'rare' mutations) and may result in different phenotypes in the same population. Just as specific variants can be associated with disease conditions, they may also cause an alteration in protein function that may be beneficial or detrimental for complex phenotypes such as health and athletic performance. In studies that assess the association of those variants with athletic ability, the genotype frequency of a group of elite athletes ('cases') is compared with the frequency in a population of sedentary non-athletes ('controls'). If one genotype is more related to elite athletic performance than the other genotypes, it will be found at a higher or lower frequency in elite athletes than in controls [12]. Several genetic variants have been found to be associated with elite performance in the last few years, and the current paradigm is that elite performance is a polygenic trait (i.e. influenced by many gene variants), with minor contributions of each variant to the unique athletic phenotype [13, 14].

Using predominately hypothesis-driven case-control candidate gene association studies, over 200 polymorphisms associated with exercise-related traits and over 20 polymorphisms associated with elite athletic status have been reported in the literature and are summarized on a yearly basis [15]. These polymorphisms are extensively discussed in the papers by Venezia and Roth [this vol., pp. 29–40] and Ahmetov et al. [this vol., pp. 41–54]. Of the polymorphisms associated with elite performance, the *ACTN3* R577X polymorphism provides the most consistent results. *ACTN3* (coding the α-actinin-3 protein) is the only gene that shows a genotype and performance association across multiple cohorts of elite athletes [16, 17, 74], and this association is strongly supported by mechanistic data from an Actn3 knockout (KO) mouse model. The angiotensin-1-converting enzyme insertion/deletion polymorphism (*ACE* I/D) is the next promising polymorphism to be associated with elite athletic performance, but its association with elite performance is less consistent than the *ACTN3* [18] and is yet to be supported by human/animal invasive (in vitro) models.

Biological Studies Can Provide Insights into the Effect of Gene Variants on Performance: The Actn3 Knockout Mouse Model as an Example

A genetically engineered mouse, in which researchers have inactivated (knocked out) or overactivated (overexpressed) a certain gene, can serve as a tool for understanding the underlying mechanisms of skeletal muscle adaptations to exercise training. To gain a greater understanding of the effects of the *ACTN3* R577X polymorphism on physiological and metabolic function at baseline (before training) and in response to exercise training, an Actn3 KO mouse model has been developed, where KO mice that are completely deficient in the α-actinin-3 protein are compared to wild-type (WT) mice, who fully express the α-actinin-3 protein [19]. Compared with WT mice, Actn3 KO mice (that mimic the *ACTN3* 577XX genotype in humans) (1) have reduced muscle mass due to decreased diameter of fast muscle fibres (where α-actinin-3 is primarily expressed), (2) show a significant decrease in grip strength, and (3) have increased endurance capacity, with KO mice running 33% further than their WT littermates at baseline [19, 20]. The absence of α-actinin-3 protein results in a shift of muscle properties towards slower muscle fibre. Fast muscle fibres (type 2/X) from KO mice have a significantly decreased anaerobic enzyme activity and an increased oxidative/mitochondrial enzyme activity, without a shift in fibre type distribution. Isolated KO muscles have longer twitch half-relaxation times and enhanced recovery from fatigue compared to the WT [19, 20]. Thus, the phenotypes of the Actn3 KO mouse mirror the gene association studies performed in humans and provide a plausible explanation for the reduced sprint performance and possible improved endurance performance in humans with the *ACTN3* 577XX genotype [21]. The shift in muscle metabolism towards oxidative metabolism could also account for the positive selection of the *ACTN3* 577X allele during human evolution; increased metabolic efficiency would increase tolerance to famine, and enhanced endurance capabilities may have conferred a survival advantage for hunter/gatherer societies. Recently, the calcineurin-specific cell-signalling pathway was suggested for improved fatigue resistance and altered muscle metabolism. It was shown that calcineurin (a calcium-dependent protein phosphatase) activity was increased ~1.9-fold in Actn3 KO mice, compared to their WT littermates [22]. Collectively, the data from the Actn3 KO model provide an example of how scientists can provide a biological explanation for the hypothesized effects of gene variants on athletic performance.

As opposed to the hypothesis-driven approach of looking at one or a few targeted polymorphisms and assessing their association with athletic ability, an 'unbiased', whole-genome sequencing approach has recently come into use. In the last few years, the cost of using whole-genome sequencing methods has dramatically dropped and become more affordable, from the first complete human genome costing USD 3 billion to sequence in 2000 to ~USD 1,000 per genome today. In addition to whole-genome sequencing, advances in molecular technologies have also enabled researchers to apply genome-wide association study (GWAS) approaches to the field. The GWAS examines the association of genetic variation with outcomes or phenotypes of interest by analysing 100,000 to several millions of polymorphisms across the entire genome, without any predetermined hypotheses about potential mechanisms. The GWAS has been successful in, among others, identifying novel genetic variants for age-related muscular degeneration [23] and type 2 diabetes mellitus [24]. A GWAS of elite human athletic performance is ongoing, but there are no published papers to date.

Tightly controlled exercise training studies are also required to understand the gene × environment effect on the adaptive response to training. Perhaps the most prominent training study is the HERITAGE study, which has identified some individuals as 'low responders' (with limited improved fitness following exercise training) and others as 'high responders' to similar exercise training, in an attempt to discover the gene variants that contribute to this phenomenon. Recently, based on data from the HERITAGE study, 2 GWASs have been published, both using $VO_{2\,max}$ as an exercise response trait [25, 26]. In the first study, the authors identified genes associated with $VO_{2\,max}$ training response, based on global skeletal muscle gene expression profiling and DNA markers [25]. A total of 29 transcripts were strongly associated with the gains in $VO_{2\,max}$, with 11 polymorphisms explaining about 23% of the variance in the $VO_{2\,max}$ training response. In the second study, a total of 39 different polymorphisms were associated with the $VO_{2\,max}$ training response, with the strongest evidence of association observed in the first intron of the acyl-CoA synthetase long-chain family member 1 *(ACSL1)* gene [26]. Nine polymorphisms explained at least 2% of the variance, while 7 contributed between 1 and 2% each. Together, 16 polymorphisms accounted for 45% of the variance in $VO_{2\,max}$ trainability, a value comparable to the heritability estimate of 47% reported previously in the HERITAGE study [25].

The Gene SMART (Skeletal Muscle Adaptive Response to Training) study, established by our research group, is an ongoing effort to use a system biology approach to identify the genomics (gene variants), transcriptomics (gene expression profile), metabolomics, and proteomics (protein abundance) that predict the response to a single bout and to 4 weeks of high-intensity interval training. We have currently recruited 40 volunteers who completed the training phase, and the goal is to recruit a total of 250 moderately trained, healthy participants (all males 18–45 years, body mass index ≤30). Participants are undergoing exercise testing and exercise training in three different centres (Victoria University, Australia; Bond University, Australia; University of Sao Paulo, Brazil), using a similar exercise programme and a tightly controlled nutrition programme, with the aim of eliminating any environmental effect on the training response, and to be able to attribute the training-induced changes to the participants' genes. Skeletal muscle biopsies and blood samples are being collected before, immediately after, and 3 h after the first training session, as well as after the 4 weeks of training. It is anticipated that this research will contribute significant new information on the genetic basis of adaptation to exercise training and will directly affect the elite performance environment.

Nurture's Influence on Athletic Ability

Besides nature, nurture (environment) also plays a major role in athletic ability. Five environmental variables are discussed in this section – deliberate practice (DP), family support, the coach's influence, relative age effect (RAE), and birthplace effect (BPE). These variables have been found to be associated with achieving a level of expertise in sport.

Deliberate Practice

DP is one of the most studied environmental factors determining expertise in sport [27–29]. DP has been defined as an activity aimed to enhance specific performance components by providing feedback and furnishing opportunities for gradual refinement of skills – or their components – through the repetitious practice most suited for attaining the desired changes in performance and its mediating mechanisms [3, 30]. DP is conceptualized as an activity performed with full concentration and is motivated by specific goals for enhancing performance. It is not the most inherently enjoyable of the available activities, and, unlike work, does not lead to immediate social or monetary rewards. In essence, DP is considered a task-specific, structured training activity that plays a major role in capturing the processes by which a skill is acquired [3].

It has been consistently argued that individual differences in performance largely reflect the difference in the amount of DP performed [31]; that is, those who have achieved the highest level of human performance have devoted more time to activities specifically designed to enhance their performance in a given domain [32–36]. In other words, extensive DP is a major determinant of attaining expertise in any given domain.

Based on a series of studies on DP in athletic performance in both individual (e.g. gymnastics, swimming, and wrestling) and team (e.g., basketball, hockey, and soccer) sports, a number of observations have been made: (a) DP features, as described below, were more evident in athletes who met their goals and became highly skilled in their specific sport domain than in those who failed to reach the expertise stage [37–41]. Among the features of DP that were found in these studies to be associated with skilled athletes were: (i) expert performers in sport accumulated more hours of DP than non-expert performers (e.g. international athletes accumulated 8,000–9,000 DP hours compared with national players who accumulated 7,500 h and provincial players who accumulated 5,000 h); (ii) experts had more organized training and individual instruction with their coaches than non-experts, and (iii) experts devoted more time to off-court/field activities, such as imagery and watching videos of their own or others' performances, than non-experts. (b) A variety of activities related to DP were reported by the athletes participating in the reviewed studies, among them on-court/field activities, learning decision-making skills, using imagery, and recreational activities. (c) A considerable variability in accumulated practice time over the years was observed in some of the studies, and therefore this variability must be addressed when examining aspects of DP. (d) Accumulated DP time increased throughout the athlete's developmental stages. It was observed that athletes devoted less time to activities associated with DP in early stages of talent development compared with the time they devoted to DP activities in the stages of specialization and expertise.

Family Support

Individuals who have achieved the highest level of proficiency in their domain (e.g. athletes, musicians, and scientists) reported that in early stages of talent development their family played a major role in encouraging and supporting their involvement in training and competition [42]. Côté [43] emphasized the importance of the family in early stages of sport involvement, particularly in the sampling stage, where children are involved in a number of sport activities. In this early stage of development, it is expected that the parents will be part of the children's sport involvement, for example by enrolling them in different sport activities, by arranging transportation and access to the sport facilities, and by providing them with some basic instruction. In a number of retrospective studies [40, 43], athletes who achieved a state of expertise in their selected sports emphasized that they could not have fulfilled their athletic potential and become the best in their domain without the involvement of their parents and siblings.

Fraser-Thomas et al. [44] offered a number of recommendations for parents who aim at helping their children to achieve, by showing them how to play a positive role in the young athlete's journey to excellence. According to the authors, parents should (a) create task-oriented climates that are autonomy supportive, (b) be particularly cautious of pushing their children, (c) teach and model life skills and positive sport participation, (d) consider how their involvement with the child-athlete influences other siblings, and (e) be aware of gender differences in their interactions with son and daughter child-athletes.

The Coach's Influence

While family members are expected to take the leading role in the young athlete's early stages of sport development, the coach takes over the role of the leading figure in advanced stages of talent development, serving as the professional authority for any aspect of the young athlete's training programme. The ability of the coach to create an environment that fosters optimal learning is one of the most significant keys to athlete development [45]. Coaches play an important role in optimizing the athletes' practice time, in some cases having optimal control over the entire training programme [46]. According to leading experts in the theory of training [47], the structure and content of practice of those athletes who achieved the highest level of proficiency in their sport is superior to that of the athletes who did not reach the same level of performance.

While working with athletes in their early stages of talent development, coaches should provide them with high levels of instruction, support and encouragement, and management behaviours with an overall positive focus [48]. Based on the recommendations, coaches should create a learning climate where the young athlete will enjoy the sport activities, be motivated to become part of the sport programme, and perform without fear of making errors. Coaches should also praise the athlete for making the required effort and provide him or her with the relevant guidance if he or she has failed to appropriately perform the learned task.

As athletes advance in their skill development, coaches can focus on the technical-tactical aspect of training by using the current knowledge from different areas of sport and exercise sciences, among them biomechanics, motor learning, exercise physiology, sport and exercise psychology, and measurement and evaluation [47, 49]. During their journey to reaching the highest level of sport performance, athletes will benefit most from training programmes in which knowledge from the above-mentioned sport sciences is synthesized. The better the synthesis among these bodies of knowledge, the higher the chances for the athlete to be prepared for practice and competition.

Relative Age Effect

The RAE implies that individuals who are relatively older than their peers in a given cohort/year (typically those who are born in the first quartile) or whose birth date is closer to the cut-off date for their age group classification, are more likely to reach higher athletic achievements [50, 51]. In a series of studies on elite male individual (e.g. swimming and tennis) and team (e.g. baseball, basketball, and soccer) sports, it was consistently found that those athletes who were born early in the competition year had a higher representation than those who were born late in that year [52, 53]. Among the reasons suggested by researchers to explain the existence of the RAE in male elite performers in sport are: (a) the older athletes are probably more experienced than the younger athletes in various motor-physical abilities, such as balance, coordination, speed, and strength, and therefore they perform these sport skills better, and (b) the older athletes in their cohort/year are more likely to be selected to better teams, and therefore are provided with guidance and training superior to that given to the younger athletes, who practice with lower-skill level athletes and may be exposed to lower-quality training conditions [50].

Concerning the RAE in female athletes, studies on basketball [54] and ice hockey [55] players showed that the RAE did exist; however, this effect reflected an overrepresentation of players in the second birth quartile compared to the overrepresentation in the first quartile found in male athletes. It was assumed by the authors of these studies that the lower rate of participation of females in sports and the lack of competition among them, compared with the higher rate of participation of males in sports and the emphasis of competition in male sports, were the reasons for the results obtained on the RAE in female athletes.

Birthplace Effect

Another environmental factor associated with achieving a high level of proficiency in sport is the place of birth – the BPE, where he or she was born, and/or where he or she grew up during the developmental years, namely up to about 14 years of age [56]. Mixed results have been reported for the BPE. For example, early studies on the BPE [52, 56] showed that male athletes who were born in cities of less than 500,000 people were more likely to play for professional leagues and attain a high level of sport competition than athletes born in larger cities. More recent studies showed different results: larger cities were just as effective in producing elite athletes as were smaller cities [53, 57].

Researchers reported that certain developmental opportunities exist that may be superior in smaller cities, such as (a) a greater amount of independent mobility and physical safety, (b) an integrative approach to sport participation involving schools, families, and the community at large, and (c) a more personal relationship between athletes and coaches [46, 56]. These conditions may provide young athletes with the most favourable conditions for developing their abilities and skills. However, it has also been argued that large cities can provide children with enhanced conditions, such as better-designed and better-equipped sporting facilities (e.g. arenas, fields, and swimming pools) and more experienced coaches.

The Missing Link – The Need for a Theory

In order to further increase our understanding of the contribution of task enhancement environmental variables to athletic ability, additional research is needed on each of the discussed variables, as well as on the interaction between these variables. For example, while reviewing the literature on the contribution of the coach to the young athlete's development, it can be observed that the amount of available research that includes children under the age of 14 appears to be relatively limited compared to the body of literature using samples consisting entirely of athletes aged 14 years and above [48]. Therefore, additional studies are needed to explore the contribution of the coach to the young athlete's ability at the initial phases of sport involvement (e.g. from ages 8 to 14). Considering the influence of the family on children's involvement in sport, future work should acknowledge and prioritize the dynamic and changing roles of parental involvement throughout the child-athletes' development, as the enormity of the changes that occur from the child's sport initiation through adolescent levels of expertise cannot be underestimated [44].

While investigating each of the discussed variables, researchers should also examine the interaction that occurs between these variables. For example, the relationships between the parents of the young athlete and his or her coach should be studied throughout the different stages of talent development. In order to provide the young athlete with the optimal preparation – physical, psychological, and emotional – both parents and coaches should be aware of the needs of the athletes and cooperate with each other in helping them cope with the multifaceted challenges they may face during their long-term athletic journey. Therefore, more information is needed on how to establish effective cooperation between parents and coaches, which ultimately will help the athletes advance from one stage of athletic development to another. In essence, the conduction of additional quantitative and qualitative inquiries will result in a better understanding of the contribution of environmental variables to athletic ability.

Fetal Programming and Epigenetics: An Example of How Our Environment Can Influence Our Genes

Fetal programming, also known as developmental programming or the Barker hypothesis, pertains to the fetal origins of adult diseases. This hypothesis is based on epidemiological data showing that low birth weight due to maternal malnutrition has long-term effects on adult health [58]. The fetal programming hypothesis suggests that the alteration in the uterine environment as a result of nutritional stress at certain stages of conceptus growth and development might permanently change tissue structure and function [59]. The offspring of malnourished mothers have a higher chance of developing coronary heart disease, stroke, diabetes, and hypertension [60]. The prevalence of type 2 diabetes increases 3-fold for men who weighed 5.5 lb or under at birth when compared to those who had a birth weight of at least 9.5 lb [61]. The correlation between human maternal nutrition and offspring birth weight, and the subsequent development of hypertension, insulin resistance, type 2 diabetes, and cardiovascular diseases, has been well studied by many researchers [61–63].

Animal studies have shown that both maternal undernutrition [64] and overnutrition [65, 66] during gestation affect skeletal muscle development. Low birth weight is associated with decreased voluntary physical activity in adult rats [67, 68]. In humans, low birth weight is associated with higher adult adiposity and decreased lean body mass [69]. Overall, these animal and human studies suggest that the environment can affect our genes, and traits such as athletic ability not only depend on the genetic code, but also on epigenetics [70, 71].

Epigenetics describes heritable changes in gene function without changes in gene sequence [72]. Epigenetic changes can pass from one cell generation to the next (mitotic inheritance), as well as from one generation of a species to the next (meiotic inheritance) [73]. Epigenetic modifications include DNA methylation, histone modification, and microRNAs, which explain the cell differentiation into different cell types with various phenotypes despite their having the same DNA sequence [73]. Interestingly, epigenetic status can be influenced by environmental factors, among them nutrition and exercise training [70]. A recent study showed that maternal overnutrition during gestation led to different epigenetic profiles in fetal skeletal muscle in animals [71].

Conclusions

We conclude that athletic ability is influenced by both nature and nurture, and with recent progress in sports genomics research into what makes an elite athlete, using the 'nature versus nurture' argument is no longer applicable. We note that the identification of novel genetic variants associated with performance is only the 'tip of the iceberg' in this field, and that the implications of training and environment must also be considered. While research in this area has progressed in the last few years, the greatest challenge will be to combine these two fields together and create models that will take into account both genetics and environmental factors. Achieving this would potentially assist in the development of a meaningful tool for coaches and exercise scientists to use in order to personalize exercise training at the highest standard.

References

1 Eynon N, Ruiz JR, Oliveira J, Duarte JA, Birk R, Lucia A: Genes and elite athletes: a roadmap for future research. J Physiol 2011;589:3063–3070.

2 Gladwell M: Outliers: The Story of Success, ed 1. New York, Little, Brown & Co, 2008.

3 Ericsson KA: Deliberate practice and the modifiability of body and mind: toward a science of the structure and acquisition of expert and elite performance. Int J Sport Psychol 2007;38:4–34.

4 Klissouras V: Heritability of adaptive variation. J Appl Physiol 1971;31:338–344.

5 Stubbe JH, Boomsma DI, Vink JM, Cornes BK, Martin NG, Skytthe A, Kyvik KO, Rose RJ, Kujala UM, Kaprio J, et al: Genetic influences on exercise participation in 37,051 twin pairs from seven countries. PLoS One 2006;1:e22.

6 De Moor MH, Spector TD, Cherkas LF, Falchi M, Hottenga JJ, Boomsma DI, De Geus EJ: Genome-wide linkage scan for athlete status in 700 British female DZ twin pairs. Twin Res Hum Genet 2007;10:812–820.

7 Bouchard C, Leon AS, Rao DC, Skinner JS, Wilmore JH, Gagnon J: The HERITAGE family study. Aims, design, and measurement protocol. Med Sci Sports Exerc 1995;27:721–729.

8 Bouchard C, An P, Rice T, Skinner JS, Wilmore JH, Gagnon J, Perusse L, Leon AS, Rao DC: Familial aggregation of VO(2max) response to exercise training: results from the HERITAGE family study. J Appl Physiol (1985) 1999;87:1003–1008.

9 An P, Perusse L, Rankinen T, Borecki IB, Gagnon J, Leon AS, Skinner JS, Wilmore JH, Bouchard C, Rao DC: Familial aggregation of exercise heart rate and blood pressure in response to 20 weeks of endurance training: the HERITAGE family study. Int J Sports Med 2003;24:57–62.

10 Perusse L, Gagnon J, Province MA, Rao DC, Wilmore JH, Leon AS, Bouchard C, Skinner JS: Familial aggregation of submaximal aerobic performance in the HERITAGE family study. Med Sci Sports Exerc 2001;33:597–604.

11 Peeters MW, Thomis MA, Beunen GP, Malina RM: Genetics and sports: an overview of the pre-molecular biology era. Med Sport Sci 2009;54:28–42.

12 Pitsiladis Y, Wang G, Wolfarth B, Scott R, Fuku N, Mikami E, He Z, Fiuza-Luces C, Eynon N, Lucia A: Genomics of elite sporting performance: what little we know and necessary advances. Br J Sports Med 2013;47:550–555.

13 Williams AG, Folland JP: Similarity of polygenic profiles limits the potential for elite human physical performance. J Physiol 2008;586:113–121.

14 Hughes DC, Day SH, Ahmetov II, Williams AG: Genetics of muscle strength and power: polygenic profile similarity limits skeletal muscle performance. J Sports Sci 2011;29:1425–1434.

15 Loos RJ, Hagberg JM, Perusse L, Roth SM, Sarzynski MA, Wolfarth B, Rankinen T, Bouchard C: Advances in exercise, fitness, and performance genomics in 2014. Med Sci Sports Exerc 2015;47:1105–1112.

16 Eynon N, Hanson ED, Lucia A, Houweling PJ, Garton F, North KN, Bishop DJ: Genes for elite power and sprint performance: *ACTN3* leads the way. Sports Med 2013;43:803–817.

17 Eynon N, Banting LK, Ruiz JR, Cieszczyk P, Dyatlov DA, Maciejewska-Karlowska A, Sawczuk M, Pushkarev VP, Kulikov LM, Pushkarev ED, et al: *ACTN3* R577X polymorphism and team-sport performance: a study involving three European cohorts. J Sci Med Sport 2014;17:102–106.

18 Puthucheary Z, Skipworth JR, Rawal J, Loosemore M, Van Someren K, Montgomery HE: The *ACE* gene and human performance: 12 years on. Sports Med 2011;41:433–448.

19 MacArthur DG, Seto JT, Raftery JM, Quinlan KG, Huttley GA, Hook JW, Lemckert FA, Kee AJ, Edwards MR, Berman Y, et al: Loss of *ACTN3* gene function alters mouse muscle metabolism and shows evidence of positive selection in humans. Nat Genet 2007;39:1261–1265.

20 MacArthur DG, Seto JT, Chan S, Quinlan KG, Raftery JM, Turner N, Nicholson MD, Kee AJ, Hardeman EC, Gunning PW, et al: An Actn3 knockout mouse provides mechanistic insights into the association between alpha-actinin-3 deficiency and human athletic performance. Hum Mol Genet 2008;17: 1076–1086.

21 Berman Y, North KN: A gene for speed: the emerging role of alpha-actinin-3 in muscle metabolism. Physiology (Bethesda) 2010;25:250–259.

22 Seto JT, Quinlan KG, Lek M, Zheng XF, Garton F, Macarthur DG, Hogarth MW, Houweling PJ, Gregorevic P, Turner N, et al: *ACTN3* genotype influences muscle performance through the regulation of calcineurin signaling. J Clin Invest 2013;123:4255–4263.

23 Swaroop A, Branham KE, Chen W, Abecasis G: Genetic susceptibility to age-related macular degeneration: a paradigm for dissecting complex disease traits. Hum Mol Genet 2007;16(spec No 2):R174–R182.

24 Stranger BE, Stahl EA, Raj T: Progress and promise of genome-wide association studies for human complex trait genetics. Genetics 2011;187:367–383.

25 Timmons JA, Knudsen S, Rankinen T, Koch LG, Sarzynski M, Jensen T, Keller P, Scheele C, Vollaard NB, Nielsen S, et al: Using molecular classification to predict gains in maximal aerobic capacity following endurance exercise training in humans. J Appl Physiol (1985) 2010;108:1487–1496.

26 Bouchard C, Sarzynski MA, Rice TK, Kraus WE, Church TS, Sung YJ, Rao DC, Rankinen T: Genomic predictors of the maximal O uptake response to standardized exercise training programs. J Appl Physiol (1985) 2011;110:1160–1170.

27 Macnamara BN, Hambrick DZ, Oswald FL: Deliberate practice and performance in music, games, sports, education, and professions: a meta-analysis. Psychol Sci 2014;25:1608–1618.

28 Baker J, Young B: 20 years later: deliberate practice and the development of expertise in sport. Int Rev Sport Exerc Psychol 2014;7:135–157.

29 Platz F, Kopiez R, Lehmann AC, Wolf A: The influence of deliberate practice on musical achievement: a meta-analysis. Front Psychol 2014;5:646.

30 Ericsson KA, Krampe RT, Teschromer C: The role of deliberate practice in the acquisition of expert performance. Psychol Rev 1993;100:363–406.

31 Ericsson KA: Why expert performance is special and cannot be extrapolated from studies of performance in the general population: a response to criticisms. Intelligence 2014;45:81–103.

32 Baker J, Cote J, Deakin J: Expertise in ultra-endurance triathletes early sport involvement, training structure, and the theory of deliberate practice. J Appl Sport Psychol 2005;17:64–78.

33 Coughlan EK, Williams AM, McRobert AP, Ford PR: How experts practice: a novel test of deliberate practice theory. J Exp Psychol Learn 2014;40:449–458.

34 Duffy LJ, Baluch B, Ericsson KA: Dart performance as a function of facets of practice amongst professional and amateur men and women players. Int J Sport Psychol 2004;35:232–245.

35 Helsen WF, Starkes JL, Hodges NJ: Team sports and the theory of deliberate practice. J Sport Exerc Psychol 1998;20:12–34.
36 Nordin SM, Cumming J, Vincent J, McGrory S: Mental practice or spontaneous play? Examining which types of imagery constitute deliberate practice in sport. J Appl Sport Psychol 2006;18:345–362.
37 Baker J, Cote J, Abernethy B: Sport-specific practice and the development of expert decision-making in team ball sports. J Appl Sport Psychol 2003;15:12–25.
38 Cumming J, Hall C: Deliberate imagery practice: the development of imagery skills in competitive athletes. J Sports Sci 2002;20:137–145.
39 Hodges NJ, Starkes JL: Wrestling with the nature of expertise: a sport specific test of Ericsson, Krampe and Tesch-Romer's (1993) theory of 'deliberate practice'. Int J Sport Psychol 1996;27:400–424.
40 Lidor R, Lavyan NZ: A retrospective picture of early sport experiences among elite and near-elite Israeli athletes: developmental and psychological perspectives. Int J Sport Psychol 2002;33:269–289.
41 Memmert D, Baker J, Bertsch C: Play and practice in the development of sport-specific creativity in team ball sports. High Abil Stud 2010;21:3–18.
42 Cutietta RA: Reference repository on Developing Talent in Young People, edited by BS Bloom, New York, Ballantine Books, 1985. Music Educ J 1985;72: 58–59.
43 Côté J: The influence of the family in the development of talent in sport. Sport Psychol 1999;13:395–417.
44 Fraser-Thomas JL, Strachan L, Jeffery-Tosoni SM: Family influence on children's involvement in sport; in Côté J, Lidor R (eds): Conditions of Children's Talent Development in Sport. Morgantown, Fitness Information Technology, 2013.
45 Baker J, Horton S, Robertson-Wilson J, Wall M: Nurturing sport expertise: factors influencing the development of elite athlete. J Sports Sci Med 2003; 2:1–9.
46 Davids K, Baker J: Genes, environment and sport performance – why the nature-nurture dualism is no longer relevant. Sports Med 2007;37:961–980.
47 Bompa TO, Buzzichelli C: Periodization training for sports, ed 3. Champaign: Human Kinetics, 2014.
48 Erickson K, Gilbert W: Coach-athlete interactions in children's sport; in Côté J, Lidor R (eds): Conditions of Children's Talent Development in Sport. Morgantown, Fitness Information Technology, 2013, pp 139–156.
49 Lidor R, Blumenstein B, Tenenbaum G: Psychological aspects of training in European basketball: conceptualization, periodization, and planning. Sport Psychol 2007;21:353–367.
50 Cobley S, Baker J, Wattie N, McKenna J: Annual age-grouping and athlete development: a meta-analytical review of relative age effects in sport. Sports Med 2009;39:235–256.
51 Wattie N, Cobley S, Baker J: Towards a unified understanding of relative age effects. J Sports Sci 2008; 26:1403–1409.
52 Baker J, Logan AJ: Developmental contexts and sporting success: birth date and birthplace effects in national hockey league draftees 2000–2005. Br J Sports Med 2007;41:515–517.
53 Schorer J, Cobley S, Busch D, Brautigam H, Baker J: Influences of competition level, gender, player nationality, career stage and playing position on relative age effects. Scand J Med Sci Sports 2009;19:720–730.
54 Delorme N, Chalabaev A, Raspaud M: Relative age is associated with sport dropout: evidence from youth categories of French basketball. Scand J Med Sci Sports 2011;21:120–128.
55 Weir PL, Smith KL, Paterson C, Horton S: Canadian Women's ice hockey – evidence of a relative age effect. Talent Dev Excell 2010;2:209–217.
56 Côté J, Macdonald DJ, Baker J, Abernethy B: When 'where' is more important than 'when': birthplace and birthdate effects on the achievement of sporting expertise. J Sports Sci 2006;24:1065–1073.
57 Schorer J, Baker J, Lotz S, Busch D: Influence of early environmental constraints on achievement motivation in talented young handball players. Int J Sport Psychol 2010;41:42–57.
58 Barker DJ, Eriksson JG, Forsen T, Osmond C: Fetal origins of adult disease: strength of effects and biological basis. Int J Epidemiol 2002;31:1235–1239.
59 Drake AJ, Walker BR: The intergenerational effects of fetal programming: non-genomic mechanisms for the inheritance of low birth weight and cardiovascular risk. J Endocrinol 2004;180:1–16.
60 Barker DJ: Coronary heart disease: a disorder of growth. Horm Res 2003;59(suppl 1):35–41.
61 Hales CN, Barker DJ, Clark PM, Cox LJ, Fall C, Osmond C, Winter PD: Fetal and infant growth and impaired glucose tolerance at age 64. BMJ 1991;303: 1019–1022.
62 Barker DJ, Osmond C: Low birth weight and hypertension. BMJ 1988;297:134–135.
63 Eriksson JG, Forsen T, Tuomilehto J, Winter PD, Osmond C, Barker DJ: Catch-up growth in childhood and death from coronary heart disease: longitudinal study. BMJ 1999;318:427–431.
64 Zhu MJ, Ford SP, Nathanielsz PW, Du M: Effect of maternal nutrient restriction in sheep on the development of fetal skeletal muscle. Biol Reprod 2004;71: 1968–1973.

65 Zhu MJ, Han B, Tong J, Ma C, Kimzey JM, Underwood KR, Xiao Y, Hess BW, Ford SP, Nathanielsz PW, et al: AMP-activated protein kinase signalling pathways are down regulated and skeletal muscle development impaired in fetuses of obese, over-nourished sheep. J Physiol 2008;586:2651–2664.
66 Tong JF, Yan X, Zhu MJ, Ford SP, Nathanielsz PW, Du M: Maternal obesity downregulates myogenesis and beta-catenin signaling in fetal skeletal muscle. Am J Physiol Endocrinol Metab 2009;296:E917–E924.
67 Vickers MH, Breier BH, Cutfield WS, Hofman PL, Gluckman PD: Fetal origins of hyperphagia, obesity, and hypertension and postnatal amplification by hypercaloric nutrition. Am J Physiol Endocrinol Metab 2000;279:E83–E87.
68 Vickers MH, Breier BH, McCarthy D, Gluckman PD: Sedentary behavior during postnatal life is determined by the prenatal environment and exacerbated by postnatal hypercaloric nutrition. Am J Physiol Regul Integr Comp Physiol 2003;285:R271–R273.
69 Kahn HS, Narayan KM, Williamson DF, Valdez R: Relation of birth weight to lean and fat thigh tissue in young men. Int J Obes Relat Metab Disord 2000;24:667–672.
70 Voisin S, Eynon N, Yan X, Bishop DJ: Exercise training and DNA methylation in humans. Acta Physiol (Oxf) 2015;213:39–59.
71 Yan X, Huang Y, Zhao JX, Rogers CJ, Zhu MJ, Ford SP, Nathanielsz PW, Du M: Maternal obesity downregulates microRNA let-7g expression, a possible mechanism for enhanced adipogenesis during ovine fetal skeletal muscle development. Int J Obes (Lond) 2013;37:568–575.
72 Bird A: Perceptions of epigenetics. Nature 2007;447:396–398.
73 Ling C, Groop L: Epigenetics: a molecular link between environmental factors and type 2 diabetes. Diabetes 2009;58:2718–2725.
74 Papadimitriou ID, Lucia A, Pitsiladis YP, Pushkarev VP, Dyatlov DA, Orekhov EF, Artioli GG, Guilherme JP, Lancha AH Jr, Ginevičienė V, Cieszczyk P, et al: ACTN3 R577X and ACE I/D gene variants influence performance in elite sprinters: a multi-cohort study. BMC Genomics 2016;17:285.

Nir Eynon, PhD
Institute of Sport, Exercise and Active Living (ISEAL)
Victoria University, PO Box 14428
Melbourne, VIC 8001 (Australia)
E-Mail Nir.Eynon@vu.edu.au

Posthumus M, Collins M (eds): Genetics and Sports, ed 2, revised, extended.
Med Sport Sci. Basel, Karger, 2016, vol 61, pp 29–40 (DOI: 10.1159/000445239)

Recent Research in the Genetics of Exercise Training Adaptation

Andrew C. Venezia · Stephen M. Roth

Department of Kinesiology, School of Public Health, University of Maryland, College Park, Md., USA

Abstract

In the present review, we focus on evaluating the state of the literature in the area of genetic aspects of exercise training adaptation, in particular focusing on findings published since 2009. Our focus is primarily on studies examining genetic polymorphisms and their association with variability in training responses for a number of exercise-related traits, including aerobic fitness, hemodynamic variables, metabolic traits, body composition and obesity, and muscular strength and size. Very few large-scale studies are available to provide the quality scientific evidence needed to conclusively identify specific genetic factors that contribute to exercise training adaptations. As such, despite years of effort by a number of groups, the search continues for specific genes and combinations of genetic factors that contribute to the interindividual variability observed in phenotype adaptations to exercise training interventions.

The purpose of this review is to provide an overview of recent publications that add to the literature of the genetics of adaptations to exercise training. While there are several studies that consider the impact of genetic polymorphisms on exercise adaptation through cross-sectional studies of athletes or active individuals, here we focus narrowly on those studies that directly examine phenotype adaptations to exercise and physical activity interventions. We focus primarily on high-quality studies published during or after 2009, though certain high-quality studies before 2009 are reviewed when appropriate. In addition, though exercise training responses are of the greatest importance to this review, investigations of lifestyle interventions that include exercise or support increased physical activity (e.g. Diabetes Prevention Program, DPP; Look AHEAD) will be discussed.

Aerobic Fitness

The HERITAGE family study has provided much of the high-quality data concerning the influence of genetic variation on aerobic exercise training-induced adaptations of a variety of traits. This large-scale, multicenter investigation included parents and adult offspring from 90 white and 40 black families and examined the role of genetic variation in a number of risk factor and performance phenotypes before and after a standardized 20-week aerobic exercise training program, in addition to establishing heritability estimates for a number of traits [1]. In an early investigation from HERITAGE, a maximal heritability estimate of 47% was determined for $VO_{2\ max}$ trainability after adjustment for age and sex [2]. As expected, the data from the multicenter training program revealed a wide range of $VO_{2\ max}$ responses to the 20-week training program and found a 2.5 times greater variance between families than within families, demonstrating clear familial aggregation in the response of $VO_{2\ max}$ to aerobic exercise training.

Though a number of studies have examined single genes and polymorphisms possibly contributing to $VO_{2\ max}$ adaptation, in 2011 Bouchard et al. [3] pushed the field with the first genome-wide association study (GWAS) of the $VO_{2\ max}$ response to exercise training. Genetic and phenotype data from 3 independent standardized exercise training programs were examined, including HERITAGE, the Dose Response to Exercise (DREW) Study [4], and the Studies of Targeted Risk Reduction Intervention through Defined Exercise (STRRIDE) [5]. Focusing first on the HERITAGE white participants, the GWAS identified 39 single-nucleotide polymorphisms (SNPs) associated with $VO_{2\ max}$ adaptation to the 20-week exercise training program. A set of 21 SNPs was identified that accounted for 49% of the variance in $VO_{2\ max}$ adaptation to training, which accounts for all of the heritability of $VO_{2\ max}$ trainability [2]. A predictor score was calculated for each subject based on the 21 identified predictor SNPs with a theoretical range of 0–42. Participants with a predictor score ≥19 had a 2.7 times greater $VO_{2\ max}$ training response compared to participants with a predictor score ≤9 (604 vs. 221 ml/min). Despite these promising results, when the 6 strongest SNPs from the predictor score were examined in the HERITAGE black participants and in the DREW and STRRIDE cohorts, only the rs11715829 within *ZIC4* was associated with $VO_{2\ max}$ gains in another cohort (STRRIDE) [3]. *ZIC4* is implicated in Dandy-Walker syndrome, a congenital brain malformation, and it is uncertain why rs11715829 influences $VO_{2\ max}$ trainability. The lack of replication of the 6 strongest predictor SNPs is likely due to the differences among the studies in the exercise training programs and sample population characteristics.

An earlier investigation by Timmons et al. [6] also relied in part on HERITAGE data and used a combination of genetic polymorphism analysis and skeletal muscle gene expression data to address genetic factors underlying the adaptation of $VO_{2\ max}$ to exercise training. They used 3 exercise studies to: (1) identify an mRNA profile associated with $VO_{2\,max}$ trainability; (2) validate that mRNA profile in an independent sample that underwent a more intense exercise intervention, and (3) generate a panel of candidate SNPs to predict $VO_{2\,max}$ trainability in the HERITAGE family study white participants. Twenty-nine mRNA probe sets were identified and correlated with $VO_{2\,max}$ trainability across the 2 independent samples. The HERITAGE family study whites were then genotyped using approximately 300 SNPs identified from those 29 predictor genes. Using stepwise regression analysis, 25 SNPs from the mRNA predictor genes were included with 10 SNPs from previous candidate genes identified in the HERITAGE samples, with the results showing 11 SNPs that explained 23% of total variance in $VO_{2\,max}$ trainability. Seven of those SNPs were from the mRNA predictor genes, and 4 were from previous genetic studies in the HERITAGE samples. When the investigators added the previously identified SNPs from HERITAGE, 3 out of 4 improved the mRNA predictor model. Interestingly, none of the 11 SNPs identified by Timmons et al. were included in the 21 predictors identified by the later GWAS of Bouchard et al. [3]. However, using more relaxed statistical requirements, 4 of the 11 predictors identified by Timmons et al. [6] were associated with $VO_{2\,max}$ trainability in the 2011 GWAS [3]. Nevertheless, considerable variation exists between the findings of the 2 studies, despite having a central core of HERITAGE participants included in both.

Interestingly, even follow-up investigations within the same sample do not consistently identify the same SNPs associated with aerobic fitness trainability. Using differentially regulated mRNAs between high and low responders to 6 weeks of endurance training (similar to Timmons et al. [6]) to identify candidate genes, Keller et al. [7] identified 24 SNPs associated with $VO_{2\,max}$ trainability in the HERITAGE sample, of which zero were also found among the SNPs identified in either of the studies by Bouchard et al. [3] or Timmons et al. [6]. In the HERITAGE whites, Rice et al. [8] found that ~20% of submaximal VO_2 trainability variance was accounted for by 13 SNPs, of which only 1 *(MIPEP)* was identified in the predictive mRNA profile previously shown to be associated with $VO_{2\,max}$ trainability [6]. *MIPEP* is predominantly expressed in cells with high oxygen consumption and is important for the maturation of proteins involved in oxidative phosphorylation [9, 10].

Beyond HERITAGE, Thomaes et al. [11] identified SNP associations with changes in peak VO_2 in a group of 935 white patients with coronary artery disease from the CAREGENE cohort that underwent 3 months of exercise training (90 min/session; 2–3 days per week; cycling, running, arm ergometry, rowing). The investigators also calculated a genetic predisposition score using multiple SNPs previously reported in the literature, and that score was a significant predictor of high $VO_{2\,peak}$ responders, though none of the genes appear to match those associated with $VO_{2\,max}$ trainability in the studies of Bouchard et al. [3], Timmons et al. [6], or Keller et al. [7].

In summary, despite a number of studies using similar genetic data from HERITAGE, little consistency has been found in identifying genetic factors contributing to the adaptation of aerobic fitness to exercise training. This is likely due to the complexity of the trait and the many underlying adapting traits that contribute to a change in $VO_{2\,max}$ in response to exercise training. As such, examining intermediary or 'subphenotypes' of $VO_{2\,max}$ may be more useful in identifying contributing genetic factors. Nevertheless, these studies demonstrate the difficulty and complexity of identifying the genetic factors contributing to a heritable phenotype.

Hemodynamic Traits

There have been few high-quality studies published in recent years concerning genetic variation and hemodynamic responses to exercise training. Much of our understanding of the influence of genetic variation on trainability of hemodynamic traits again comes from the HERITAGE family study. In 2002, Rankinen et al. [12] performed an initial genome-wide linkage scan for submaximal exercise cardiac output and stroke volume response to the standardized 20-week training program and found a few loci significantly associated with the traits. Using those loci, Rankinen et al. [13] later identified 2 markers near the titan gene (*TTN*) that showed the strongest linkage with training-induced submaximal stroke volume and proposed *TTN* as a likely candidate gene for submaximal stroke volume adaptation. The TTN protein is known to play an important role in cardiac health and disease [14] making it an ideal candidate gene for cardiac adaptations to exercise training. It plays an important role in the structural integrity of cardiac muscle cells and might be a primary contributor to the Frank-Starling law [14], potentially explaining the association with training-induced submaximal stroke volume. In a more recent follow-up investigation in the HERITAGE participants, the *KIF5B* gene was also associated with submaximal stroke volume trainability. The SNPs identified in *KIF5B* were found to be functional, with influences on gene promoter activity and impacts on mitochondrial content in cell culture. Thus, both *KIF5B* and *TTN* appear to harbor genetic variation associated with stroke volume trainability.

Recent publications have also identified candidate genes and SNPs associated with submaximal heart rate trainability. Following up a genome-wide linkage scan from HERITAGE that identified loci associated with submaximal heart rate adaptation to training [15], Rankinen et al. [16] identified SNPs near the *CREB1* gene that were associated with changes in submaximal heart rate response to exercise training. The strongest associations were found for rs2253206 and rs2360969, with rs2253206 shown to impact *CREB1* promoter activity in cell culture. Rankinen et al. [17] later performed a GWAS to examine more broadly the genetic factors associated with submaximal heart rate adaptation to exercise training and identified 9 SNPs that accounted for 34% of the variance in response. Importantly, *CREB1* was among those 9 SNPs, which supports the previous finding by Rankinen et al. [16] and points to *CREB1* as a possible gene underlying heart rate adaptations to exercise training. *CREB1* is a transcription factor known to be important for transcription of genes essential for long-term memory [18], and Rankinen et al. [16] suggested a possible exercise-induced 'cardiac memory' mediated by *CREB1*.

Glucose and Insulin

Exercise training is an effective method to improve insulin sensitivity and glucose tolerance; however, there is considerable individual variability in response to exercise training [19]. Early reports from the HERITAGE family study identified chromosomal regions [20] and genetic variants [21] associated with the exercise training response of plasma insulin and glucose tolerance. Common exonic variants in the leptin and leptin receptor genes were significantly associated with insulin sensitivity and glucose tolerance response to exercise in HERITAGE whites and blacks [21]. Similarly, a polymorphism (rs180058) in the hepatic lipase gene (*LIPC*) was associated with the insulin sensitivity response to exercise training in HERITAGE whites and blacks [22], while SNPs in the *FHL1* gene, which encodes a scaffold protein with high expression in oxidative skeletal muscle fibers [23], were associated with training adaptations in glucose homeostasis, with differing results in men and women [24].

The influence of polymorphisms in the peroxisome proliferator-activated receptor γ (*PPARG*) gene that are associated with conversion from impaired glucose tolerance to type 2 diabetes was shown to be modified by participation in physical activity as part of the DPP [25]. PPARG is known to play an important role in insulin sensitivity, and rodent studies demonstrate that manipulations of the gene that alter activity strongly influence insulin sensitivity [26]. Though physical activity was assessed via questionnaire, and no specific training programs were provided in the DPP lifestyle intervention, participants received counseling on diet and exercise, and were encouraged to participate in endurance and resistance training. The effect of risk alleles in each *PPARG* SNP was reduced with physical activity, though a similar investigation from the Finnish Diabetes Prevention Study failed to replicate those findings [27]. In a similar analysis in HERITAGE, the *PPARG* Pro12Ala polymorphism was examined along with 6 other candidate genes for associations with exercise training responses in glucose homeostasis [28]. Of those genes examined, only the *PPARG* Pro12Ala polymorphism was significantly associated with glucose homeostasis traits, such that glucose tolerance, insulin response to glucose, glucose effectiveness, and disposition index responses to exercise training were greater in Ala carriers compared to Pro/Pro homozygotes. Thus, there is growing evidence showing physical activity as a modifier of the risk associated with progression to type 2 diabetes associated with the *PPARG* genotype.

Three additional reports from the DPP cohort investigated genetic variants associated with exercise training and diabetes risk. Two of these investigations identified modest associations between physical activity and modification of diabetes risk factor alleles in the *MC4R* gene [29] and the *ENPP1* gene [30]. Numerous *MC4R* mutations are associated with obesity in humans, and blocking the MC4R pathway in rodents induces obesity [31]. ENPP1 regulates insulin action by binding to and inhibiting the insulin receptor and has a clear role in insulin sensitivity and diabetes [32]. Hivert et al. [33] used a genetic risk score of 34 type 2 diabetes-associated loci to examine the risk of progression to type 2 diabetes in the DPP cohort. Though a higher risk score was associated with a higher risk of development of type 2 diabetes, the lifestyle intervention was only shown to reduce this risk in a subset of individuals with the highest genetic risk score.

In summary, little work has been done establishing the role of genetic variation in explaining the variability of insulin and glucose responses to exercise training. More work has been done examining the role of physical activity on modifying the genetic risk of progression to type 2 diabetes, with *PPARG* as a promising candidate gene in this area.

Lipids

Few recent high-quality studies have been published in the area of genetics of exercise training and lipid and lipoprotein metabolism. Two investigations by Pollin et al. [34, 35] identified genetic variants associated with baseline lipid levels in the DPP cohort, but few interactions with lifestyle intervention were identified. In the first study, the glucokinase regulatory protein gene (*GCKR*) was strongly associated with triglyceride levels and moderately associated with fasting glucose levels at baseline [34]. GCKR inhibits glucokinase activity and is important for hepatic glucose and lipid metabolism [36]. The association of the *GCKR* 466L allele with elevated triglyceride levels was reduced after the 1-year lifestyle intervention. This allele is associated with reduced stimulated glucokinase inhibitory activity [37]. In the second study [35], the researchers created a genetic risk score based on 32 previously identified risk variants for fasting lipids. They determined that the risk score was significantly associated with baseline lipid profiles in DPP but there was no significant interaction with the 1-year lifestyle intervention.

The Look AHEAD study was a large multicenter investigation with over 5,145 overweight participants with type 2 diabetes who participated in a lifestyle intervention program that encouraged increases in moderate-intensity physical activity [38]. Huggins et al. [39] investigated 82 SNPs from 31 loci identified from various GWASs focused on high-density lipoprotein cholesterol and triglyceride levels. The investigators identified 6 SNPs with nominal interactions with response to the lifestyle intervention for high-density lipoprotein cholesterol and 2 SNPs for triglycerides. Polymorphisms in the *CETP*, *LIPC*, and *PGS1* genes showed associations with lipid levels that were modified significantly by the lifestyle intervention. Each of these genes is involved in lipid metabolic pathways [40, 41].

Body Composition and Adiposity

Concerning exercise training and genetic determinants of obesity and adiposity, the majority of research has focused on the fat mass and obesity-associated gene *(FTO)*. Though other variants associated with body mass index (BMI) have recently been identified [42, 43], *FTO* remains the most extensively investigated locus for BMI and body composition traits as it relates to physical activity and exercise training [44]. The *FTO* gene has a substantial literature base supporting its association with obesity-related traits [45], and a recent meta-analysis supports the conclusion that increased physical activity reduces the influence of *FTO* risk alleles on obesity risk [46]; however, most studies assessed physical activity via questionnaire, and comparatively few studies have used exercise training interventions. Here, we will focus only on those studies that have used direct exercise training or lifestyle interventions to examine the modification of *FTO* associations with obesity risk.

In the HERITAGE family study, white A/A homozygotes of rs8050136 in *FTO* tended to have a higher BMI than C/C homozygotes, in agreement with previous research; however, in contrast to cross-sectional studies that suggest that physical activity reduces the impact of *FTO* risk genotypes on body composition, A/A homozygotes (with a higher obesity risk) showed a smaller decrease in fat mass following the 20-week exercise training program compared to C allele carriers [47]. It is important to note that many cross-sectional studies are primarily investigating prevention of weight gain over time as opposed to weight loss, which might explain this discrepancy. In the DPP cohort, the *FTO* genotype for rs9939609 (in complete linkage disequilibrium with rs8050136) was associated with BMI and body weight at baseline but was not associated with any obesity-related traits following the 1-year lifestyle intervention [48], indicating a reduction of risk as a consequence of intervention. In contrast, no association between the rs9939609 polymorphism and obesity-related traits following lifestyle intervention was observed in the Finnish Diabetes Prevention Study [49], which included 522 middle-aged and overweight participants with impaired glucose tolerance. During the 4-year follow-up period, carriers of the risk A allele had a higher BMI, as would be expected based on previous research, though there was no influence of the *FTO* genotype on weight change following the long-term lifestyle intervention.

Mitchell et al. [50] examined postmenopausal women who entered control or 1 of 3 exercise intervention groups that required 3–4 sessions per week of exercise training that would amount to 50, 100 or 150% of the NIH Consensus Development Panel recommendations for physical activity (4, 8, or 12 kcal/kg/week, respectively) for 6 months. As expected, the A/A homozygotes for *FTO* rs8050136 had a higher BMI at baseline compared to C/C homozygotes. In response to the exercise intervention, A/A homozygotes in the 100 and 150% exercise training groups exhibited a significantly larger weight loss after intervention compared to C allele carriers, consistent with the bulk of cross-sectional investigations of *FTO* genotype and interaction with physical activity.

In contrast to cross-sectional studies that demonstrate that physical activity attenuates the influence of *FTO* risk alleles on obesity traits [46], the research on the influence of the *FTO* genotype on exercise training- and lifestyle intervention-induced changes in obesity-related traits is less conclusive. Across the 4 studies described above, 2 show an interaction of physical activity attenuating the risk of *FTO* genotypes on obesity risk, 1 reported the risk allele to be less responsive to such an intervention, and 1 reported no interaction.

Muscular Adaptations

Gene association studies have identified a number of polymorphisms in genes such as *IL15* [51], *ACE* [52], *ACTN3* [53], and *IGF1* [54, 55] that are associated with muscular adaptations to strength training; however, few gene association studies have been published since 2009, and even fewer have relied on the approach of using the mRNA transcriptome to examine genetic influences for training adaptation, as was described earlier. As such, we will focus only on the most impactful of the studies in this area.

Most of the studies published since 2009 arise from the Functional Single Nucleotide Polymorphisms Associated with Human Muscle Size and Strength (FAMuSS) cohort, which focused on identifying SNPs and genetic factors associated with muscular adaptations to resistance training. The FAMuSS study included approximately 1,300 young healthy men and women who underwent muscle size and strength measures before and after a 12-week unilateral resistance training program to increase muscle size and strength of the elbow flexor and extensor muscles of the nondominant arm [56]. Participants were genotyped for approximately 500 polymorphisms with the aim of identifying variants associated with baseline and resistance training response of muscle size and strength. Though the SNPs found to be associated with either size or strength adaptations to resistance training accounted for a very small amount of overall variance, they offer insight into genes that may contribute to muscular training response. Polymorphisms in the *BMP2* gene [57], *CNTF* [58], *CCL2* and its receptor *CCR2* [59], and *SPP1* [60] have all been associated with muscular adaptations to the 12-week resistance training program; however, the influence of these SNPs is not always consistent for both men and women, nor have they been replicated in other training studies. The search for genetic factors related to the strength training adaptation of muscular traits remains in the earliest of stages.

Similar to the work described above using mRNA transcriptome profiles to identify genes of interest, Bamman and his group [61–63] have performed a series of experiments examining extreme responders, moderate responders, and nonresponders to strength training, classified by measurement of myofiber hypertrophy of the vastus lateralis muscle after 16 weeks of progressive resistance training. The first [61] and second studies [62] focused on candidate muscle hypertrophy transcripts and satellite cell activation, respectively, and found significant differences in transcription profiles among responder clusters. In the most recent investigation [63], the research group identified pretraining transcriptional profiles that differed between the clusters of responders. They observed 8,026 baseline transcripts differentially expressed between extreme and nonresponders, 2,463 between extreme and moderate responders, and 1,294 between moderate and nonresponders. The authors suggested that the extreme responders are potentially characterized by a profile of baseline mRNA transcripts 'primed' for muscular adaptation to strength training while the nonresponders have a pretraining profile primed for proinflammatory gene transcription. Like genotype, a pretraining transcriptional profile has the potential to predict the potential for training adaptations, though biopsies for such a prediction are a hurdle. The potential for SNPs in these genes to be used as a means of prediction (rather than direct biopsy) is similar to the approaches used by the Bouchard group for aerobic fitness, described above.

A similarly designed study by Phillips et al. [64] using transcriptional profiling identified over 600 mRNA transcripts that covaried with muscle hypertrophy magnitude following 20 weeks of resistance training. The gene transcripts that were associated with response to resistance training in the papers of Phillips et al. [64] and Thalacker-Mercer et al. [63] are not consistent, likely due to differences in transcriptome arrays and sample population. In summary, based on the current evidence in this area, we currently know very little about the genetic predictors of the muscular adaptations to resistance training.

Conclusion

Despite ~15 years since the sequencing of the human genome, the goal of identifying the genetic factors underlying the adaptations of various traits to exercise training remains unmet and elusive. This is in large measure due to the lack of large exercise training interventions that allow the sample sizes necessary for more conclusive gene association studies, but also to the incredible complexity of the genetic influences underlying these complex traits. The emerging use of system biology approaches to identify the source of interindividual variability, such as the mRNA transcriptome work described here and the addition of epigenetic work that is just now being completed, will no doubt add to our ability to identify influential (and perhaps predictive) genetic factors, but a key limitation compared to other fields will be the need to provide controlled, consistent, and high-quality exercise training interventions for large numbers of people. Such work will always be a challenge given the cost and logistical constraints to these kinds of efforts and should be a reminder to the field of the importance of collaboration and an eye to combining data sets of similar quality and design.

Acknowledgment

Andrew Venezia was funded by the NIH predoctoral training grant F31 MH103951.

References

1 Bouchard C, Leon AS, Rao DC, Skinner JS, Wilmore JH, Gagnon J: The HERITAGE family study. Aims, design, and measurement protocol. Med Sci Sports Exerc 1995;27:721–729.

2 Bouchard C, An P, Rice T, Skinner JS: Familial aggregation of VO_2 max response to exercise training: results from the HERITAGE family study. J Appl Physiol 1999;87:1003–1008.

3 Bouchard C, Sarzynski MA, Rice TK, Kraus WE, Church TS, Sung YJ, et al: Genomic predictors of the maximal O_2 uptake response to standardized exercise training programs. J Appl Physiol 2011;110: 1160–1170.

4 Morss GM, Jordan AN, Skinner JS, Dunn AL, Church TS, Earnest CP, et al: Dose response to exercise in women aged 45–75 years (DREW): design and rationale. Med Sci Sports Exerc 2004;36:336–344.

5 Kraus WE, Torgan CE, Duscha BD, Norris J, Brown SA, Cobb FR, et al: Studies of a targeted risk reduction intervention through defined exercise (STRRIDE). Med Sci Sports Exerc 2001;33:1774–1784.

6 Timmons JA, Knudsen S, Rankinen T, Koch LG, Sarzynski M, Jensen T, et al: Using molecular classification to predict gains in maximal aerobic capacity following endurance exercise training in humans. J Appl Physiol 2010;108:1487–1496.

7 Keller P, Vollaard NBJ, Gustafsson T, Gallagher IJ, Sundberg CJ, Rankinen T, et al: A transcriptional map of the impact of endurance exercise training on skeletal muscle phenotype. J Appl Physiol 2011;110: 46–59.

8 Rice TK, Sarzynski MA, Sung YJ, Argyropoulos G, Stütz AM, Teran-Garcia M, et al: Fine mapping of a QTL on chromosome 13 for submaximal exercise capacity training response: the HERITAGE Family Study. Eur J Appl Physiol 2011;112:2969–2978.

9 Chew A, Buck EA, Peretz S, Sirugo G, Rinaldo P, Isaya G: Cloning, expression, and chromosomal assignment of the human mitochondrial intermediate peptidase gene (MIPEP). Genomics 1997;40:493–496.

10 Chew A, Sirugo G, Alsobrook JP II, Isaya G: Functional and genomic analysis of the human mitochondrial intermediate peptidase, a putative protein partner of frataxin. Genomics 2000;65:104–112.

11 Thomaes T, Thomis M, Onkelinx S, Fagard R, Matthijs G, Buys R, et al: A genetic predisposition score for muscular endophenotypes predicts the increase in aerobic power after training: the CAREGENE study. BMC Genet 2011;12:84–84.

12 Rankinen T, An P, Pérusse L, Rice T, Chagnon YC, Gagnon J, et al: Genome-wide linkage scan for exercise stroke volume and cardiac output in the HERITAGE Family Study. Physiol Genomics 2002;10:57–62.

13 Rankinen T, Rice T, Boudreau A, Leon AS, Skinner JS, Wilmore JH, et al: Titin is a candidate gene for stroke volume response to endurance training: the HERITAGE Family Study. Physiol Genomics 2003; 15:27–33.

14 Linke WA, Hamdani N: Gigantic Business: Titin Properties and Function Through Thick and Thin. Circ Res 2014;114:1052–1068.

15 Spielmann N, Leon AS, Rao DC, Rice T, Skinner JS, Rankinen T, et al: Genome-wide linkage scan for submaximal exercise heart rate in the HERITAGE family study. Am J Physiol 2007;293:H3366–H3371.

16 Rankinen T, Argyropoulos G, Rice T, Rao DC, Bouchard C: CREB1 is a strong genetic predictor of the variation in exercise heart rate response to regular exercise: the HERITAGE family study. Circ Cardiovasc Genet 2010;3:294–299.

17 Rankinen T, Sung YJ, Sarzynski MA, Rice TK, Rao DC, Bouchard C: Heritability of submaximal exercise heart rate response to exercise training is accounted for by nine SNPs. J Appl Physiol 2012;112: 892–897.

18 Kida S, Josselyn SA, de Ortiz SP, Kogan JH, Chevere I, Masushige S, et al: CREB required for the stability of new and reactivated fear memories. Nat Neurosci 2002;5:348–355.

19 Boulé NG, Weisnagel SJ, Lakka TA, Tremblay A, Bergman RN, Rankinen T, et al: Effects of exercise training on glucose homeostasis: the HERITAGE family study. Diabetes Care 2005;28:108–114.

20 Lakka TA, Rankinen T, Weisnagel SJ, Chagnon YC, Rice T, Leon AS, et al: A quantitative trait locus on 7q31 for the changes in plasma insulin in response to exercise training: the HERITAGE family study. Diabetes 2003;52:1583–1587.

21 Lakka TA, Rankinen T, Weisnagel SJ, Chagnon YC, Lakka H-M, Ukkola O, et al: Leptin and leptin receptor gene polymorphisms and changes in glucose homeostasis in response to regular exercise in nondiabetic individuals: the HERITAGE family study. Diabetes 2004;53:1603–1608.

22 Teran-Garcia M, Santoro N, Rankinen T, Bergeron J, Rice T, Leon AS, et al: Hepatic lipase gene variant -514C>T is associated with lipoprotein and insulin sensitivity response to regular exercise: the HERITAGE family study. Diabetes 2005;54:2251–2255.

23 Loughna PT, Mason P, Bayol S, Brownson C: The LIM-domain protein FHL1 (SLIM 1) exhibits functional regulation in skeletal muscle. Mol Cell Biol Res Commun 2000;3:136–140.

24 Teran-Garcia M, Rankinen T, Rice T, Leon AS, Rao DC, Skinner JS, et al: Variations in the four and a half LIM domains 1 gene (FHL1) are associated with fasting insulin and insulin sensitivity responses to regular exercise. Diabetologia 2007;50:1858–1866.

25 Kilpeläinen TO, Lakka TA, Laaksonen DE, Lindström J, Eriksson JG, Valle TT, et al: SNPs in PPARG associate with type 2 diabetes and interact with physical activity. Med Sci Sports Exerc 2008;40:25–33.

26 Rangwala SM, Lazar MA: Peroxisome proliferator-activated receptor γ in diabetes and metabolism. Trends Pharmacol Sci 2004;25:331–336.

27 Kilpeläinen TO, Lakka TA, Laaksonen DE, Mager U, et al: Interaction of single nucleotide polymorphisms in ADRB2, ADRB3, TNF, IL6, IGF1R, LIPC, LEPR, and GHRL with physical activity on the risk of type 2 diabetes mellitus and changes in characteristics of the metabolic syndrome: the Finnish Diabetes Prevention Study. Metabolism 2008;57:428–436.

28 Ruchat SM, Rankinen T, Weisnagel SJ, Rice T, Rao DC, Bergman RN, et al: Improvements in glucose homeostasis in response to regular exercise are influenced by the PPARG Pro12Ala variant: results from the HERITAGE family study. Diabetologia 2009;53:679–689.

29 Pan Q, Delahanty LM, Jablonski KA, Knowler WC, Kahn SE, Florez JC, et al: Variation at the melanocortin 4 receptor gene and response to weight-loss interventions in the Diabetes Prevention Program. Obesity 2013;21:E520–E526.

30 Moore AF, Jablonski KA, Mason CC, McAteer JB, Arakaki RF, Goldstein BJ, et al: The association of ENPP1K121Q with diabetes incidence is abolished by lifestyle modification in the Diabetes Prevention Program. J Clin Endocrinol Metab 2009;94:449–455.

31 Kebede MA, Attie AD: Insights into obesity and diabetes at the intersection of mouse and human genetics. Trends Endocrinol Metab 2014;25:493–501.

32 Bacci S, De Cosmo S, Prudente S, Trischitta V: ENPP1 gene, insulin resistance and related clinical outcomes. Curr Opin Clin Nutr Metab Care 2007;10:403–409.

33 Hivert MF, Jablonski KA, Perreault L, Saxena R, McAteer JB, Franks PW, et al: Updated genetic score based on 34 confirmed type 2 diabetes loci is associated with diabetes incidence and regression to normoglycemia in the Diabetes Prevention Program. Diabetes 2011;60:1340–1348.

34 Pollin TI, Jablonski KA, McAteer JB, Saxena R, Kathiresan S, Kahn SE, et al: Triglyceride response to an intensive lifestyle intervention is enhanced in carriers of the GCKRPro446Leu polymorphism. J Clin Endocrinol Metab 2011;96:E1142–E1147.

35 Pollin TI, Isakova T, Jablonski KA, de Bakker PIW, Taylor A, McAteer J, et al: Genetic modulation of lipid profiles following lifestyle modification or metformin treatment: the Diabetes Prevention Program. PLoS Genet 2012;8:e1002895.

36 Raimondo A, Rees MG, Gloyn AL: Glucokinase regulatory protein. Curr Opin Lipid 2015;26:88–95.

37 Beer NL, Tribble ND, McCulloch LJ, Roos C, Johnson PRV, Orho-Melander M, et al: The P446L variant in GCKR associated with fasting plasma glucose and triglyceride levels exerts its effect through increased glucokinase activity in liver. Hum Mol Genet 2009;18:4081–4088.

38 Wadden TA, West DS, Delahanty L, Jakicic J, Rejeski J, Williamson D, et al: The Look AHEAD study: a description of the lifestyle intervention and the evidence supporting it. Obesity 2006;14:737–752.

39 Huggins GS, Papandonatos GD, Erar B, Belalcazar LM, Brautbar A, Ballantyne C, et al: Do genetic modifiers of high-density lipoprotein cholesterol and triglyceride levels also modify their response to a lifestyle intervention in the setting of obesity and type-2 diabetes mellitus? The Action for Health in Diabetes (Look AHEAD) Study. Circ Cardiovasc Genet 2013;6:391–399.

40 Willer CJ, Schmidt EM, Sengupta S, Peloso GM, Gustafsson S, Kanoni S, et al: Discovery and refinement of loci associated with lipid levels. Nat Genet 2013;45:1274–1283.

41 Teslovich TM, Musunuru K, Smith AV, Edmondson AC, Stylianou IM, Koseki M, et al: Biological, clinical and population relevance of 95 loci for blood lipids. Nature 2010;466:707–713.

42 Hoggart CJ, Venturini G, Mangino M, Gomez F, Ascari G, Zhao JH, et al: Novel approach identifies SNPs in SLC2A10 and KCNK9 with evidence for parent-of-origin effect on body mass index. PLoS Genet 2014;10:e1004508.

43 Wen W, Zheng W, Okada Y, Takeuchi F, Tabara Y, Hwang J-Y, et al: Meta-analysis of genome-wide association studies in East Asian-ancestry populations identifies four new loci for body mass index. Hum Mol Genet 2014;23:5492–5504.

44 Loos RJF, Hagberg JM, Pérusse L, Roth SM, Sarzynski MA, Wolfarth B, et al: Advances in exercise, fitness, and performance genomics in 2014. Med Sci Sports Exerc 2015;47:1105–1112.

45 Fawcett KA, Barroso I: The genetics of obesity: FTO leads the way. Trends Genet 2010;26:266–274.

46 Kilpeläinen TO, Qi L, Brage S, Sharp SJ, Sonestedt E, Demerath E, et al: Physical activity attenuates the influence of FTO variants on obesity risk: a meta-analysis of 218,166 adults and 19,268 children. PLoS Med 2011;8:e1001116.

47 Rankinen T, Rice T, Teran-Garcia M, Rao DC, Bouchard C: FTO genotype is associated with exercise training-induced changes in body composition. Obesity 2009;18:322–326.
48 Franks PW, Jablonski KA, Delahanty LM, McAteer JB, Kahn SE, Knowler WC, et al: Assessing gene-treatment interactions at the FTO and INSIG2 loci on obesity-related traits in the Diabetes Prevention Program. Diabetologia 2008;51:2214–2223.
49 Lappalainen TJ, Tolppanen A-M, Kolehmainen M, Schwab U, Lindström J, Tuomilehto J, et al: The common variant in the FTOGene did not modify the effect of lifestyle changes on body weight: the Finnish Diabetes Prevention Study. Obesity 2012;17:832–836.
50 Mitchell JA, Church TS, Rankinen T, Earnest CP, Sui X, Blair SN: FTO genotype and the weight loss benefits of moderate intensity exercise. Obesity 2009;18:641–643.
51 Riechman SE, Balasekaran G, Roth SM, Ferrell RE: Association of interleukin-15 protein and interleukin-15 receptor genetic variation with resistance exercise training responses. J Appl Physiol 2004;97:2214–2219.
52 Pescatello LS, Kostek MA, Gordish-Dressman H, Thompson PD, Seip RL, Price TB, et al: ACE ID genotype and the muscle strength and size response to unilateral resistance training. Med Sci Sports Exerc 2006;38:1074–1081.
53 Clarkson PM, Devaney JM, Gordish-Dressman H, Thompson PD, Hubal MJ, Urso M, et al: ACTN3 genotype is associated with increases in muscle strength in response to resistance training in women. J Appl Physiol 2005;99:154–163.
54 Hand BD, Kostek MC, Ferrell RE, Delmonico MJ, Douglass LW, Roth SM, et al: Influence of promoter region variants of insulin-like growth factor pathway genes on the strength-training response of muscle phenotypes in older adults. J Appl Physiol 2007;103:1678–1687.
55 Kostek MC, Delmonico MJ, Reichel JB, Roth SM, Douglass L, Ferrell RE, et al: Muscle strength response to strength training is influenced by insulin-like growth factor 1 genotype in older adults. J Appl Physiol 2005;98:2147–2154.
56 Pescatello LS, Devaney JM, Hubal MJ, Thompson PD, Hoffman EP: Highlights from the Functional Single Nucleotide Polymorphisms Associated with Human Muscle Size and Strength or FAMuSS Study. Biomed Res Int 2013;2013:643575.
57 Devaney JM, Tosi LL, Fritz DT, Gordish-Dressman HA, Jiang S, Orkunoglu-Suer FE, et al: Differences in fat and muscle mass associated with a functional human polymorphism in a post-transcriptional BMP2 gene regulatory element. J Cell Biochem 2009;107:1073–1082.
58 Walsh S, Kelsey BK, Angelopoulos TJ, Clarkson PM, Gordon PM, Moyna NM, et al: CNTF 1357 G->A polymorphism and the muscle strength response to resistance training. J Appl Physiol 2009;107:1235–1240.
59 Harmon BT, Orkunoglu-Suer EF, Adham K, Larkin JS, Gordish-Dressman H, Clarkson PM, et al: CCL2 and CCR2 variants are associated with skeletal muscle strength and change in strength with resistance training. J Appl Physiol 2010;109:1779–1785.
60 Hoffman EP, Gordish-Dressman H, McLane VD, Devaney JM, Thompson PD, Visich P, et al: Alterations in osteopontin modify muscle size in females in both humans and mice. Med Sci Sports Exerc 2013;45:1060–1068.
61 Bamman MM, Petrella JK, Kim JS, Mayhew DL, Cross JM: Cluster analysis tests the importance of myogenic gene expression during myofiber hypertrophy in humans. J Appl Physiol 2007;102:2232–2239.
62 Petrella JK, Kim JS, Mayhew DL, Cross JM, Bamman MM: Potent myofiber hypertrophy during resistance training in humans is associated with satellite cell-mediated myonuclear addition: a cluster analysis. J Appl Physiol 2008;104:1736–1742.
63 Thalacker-Mercer A, Stec M, Cui X, Cross J, Windham S, Bamman M: Cluster analysis reveals differential transcript profiles associated with resistance training-induced human skeletal muscle hypertrophy. Physiol Genomics 2013;45:499–507.
64 Phillips BE, Williams JP, Gustafsson T, Bouchard C, Rankinen T, Knudsen S, et al: Molecular networks of human muscle adaptation to exercise and age. PLoS Genet 2013;9:e1003389.

Stephen M. Roth
Department of Kinesiology
School of Public Health, University of Maryland
2242 Valley Drive, College Park, MD 20742 (USA)
E-Mail sroth1@umd.edu

Posthumus M, Collins M (eds): Genetics and Sports, ed 2, revised, extended.
Med Sport Sci. Basel, Karger, 2016, vol 61, pp 41–54 (DOI: 10.1159/000445240)

Genes and Athletic Performance: An Update

Ildus I. Ahmetov[a–c] • Emiliya S. Egorova[b] • Leysan J. Gabdrakhmanova[a] • Olga N. Fedotovskaya[d]

[a]Sport Technology Research Center, Volga Region State Academy of Physical Culture, Sport and Tourism, and [b]Laboratory of Molecular Genetics, Kazan State Medical University, Kazan, and [c]Department of Molecular Biology and Genetics, Research Institute for Physical-Chemical Medicine, Moscow, Russia; [d]Department of Physiology and Pharmacology, Karolinska Institutet, Stockholm, Sweden

Abstract

Humans vary in their ability to achieve success in sports, and this variability mostly depends on genetic factors. The main goal of this work was to review the current progress in the understanding of genetic determinism of athlete status and to describe some novel and important DNA polymorphisms that may underlie differences in the potential to be an elite athlete. In the past 19 years, at least 155 genetic markers (located within almost all chromosomes and mtDNA) were found to be linked to elite athlete status (93 endurance-related genetic markers and 62 power/strength-related genetic markers). Importantly, 41 markers were identified within the last 2 years by performing genome-wide association studies (GWASs) of African-American, Jamaican, Japanese, and Russian athletes, indicating that GWASs represent a promising and productive way to study sports-related phenotypes. Of note, 31 genetic markers have shown positive associations with athlete status in at least 2 studies and 12 of them in 3 or more studies. Conversely, the significance of 29 markers was not replicated in at least 1 study, raising the possibility that several findings might be false-positive. Future research, including multicentre GWASs and whole-genome sequencing in large cohorts of athletes with further validation and replication, will substantially contribute to the discovery of large numbers of the causal genetic variants (mutations and DNA polymorphisms) that would partly explain the heritability of athlete status and related phenotypes.

Genetic factors are considered to play a key role in athletic performance and related phenotypes such as power, strength, aerobic capacity, flexibility, coordination, and temperament. Despite a relatively high heritability of athlete status (up to ~70% depending on sport discipline) [1] and intermediate phenotypes [2, 3], the search for genetic variants contributing to predisposition to success in certain types of sport has been a challenging task. Sports genomics is a relatively new scientific discipline focusing on the organization and functioning of the genome of elite athletes [4]. The era of sports genomics began in the early 2000s with the discovery of the first genetic markers associated with athletic performance (e.g. *ACE*, *ACTN3*, *AMPD1*, and *PPARGC1A* gene polymorphisms). With genotyping, sequencing and the use of DNA microarray becoming widely available, a large number of genetic studies evaluating candidate gene variants have been published with largely unconfirmed associations with elite athlete status [4].

Case-control studies remain the most common study design in sports genomics and generally involve determining whether one allele of a DNA sequence (gene or non-coding region of DNA) is more common in a group of elite athletes than it is in the general population, thus implying that the allele boosts performance. To avoid false-positive results, case-control studies should have at least 1 replication with additional athletic and non-athletic cohorts from different populations (external replication) or subgroups of the same cohort (internal replication) [5–7]. Another way to identify sports-related genetic markers is to compare genotype and allelic frequencies between athletes with the best and the worst competition results [8, 9].

Since endurance and power are located at opposite extremes of the muscle performance continuum, the comparison of allelic or genotype frequencies between endurance and power athletes (or power/strength athletes vs. athletes involved in low-intensity sports, etc.) is also a way to identify endurance/power markers [10, 11]. Cross-sectional association studies are another type of study design in sports genomics and examine whether athletes with one genotype (or allele) of a particular DNA sequence have different measures of a trait (e.g. $VO_{2\ max}$, running time, percentage of fast-twitch muscle fibres, cardiac size, lactate, etc.) compared to the rest of the sample [12, 13].

A genome-wide association study (GWAS) is a new approach that involves rapidly scanning several hundred thousand (up to 5 million) markers across the complete sets of DNA by microchips of many people to find DNA polymorphisms associated with a particular trait. One of the advantages of the GWAS approach is that it is unbiased with respect to genomic structure and previous knowledge of the trait (hypothesis-free), in contrast to candidate gene studies, where knowledge of the trait is used to identify candidate loci contributing to the trait of interest [14]. Thus, GWASs facilitated by high-throughput genotyping technologies have been enormously successful in identifying single-nucleotide polymorphisms (SNPs) that are associated with complex traits.

DNA polymorphisms (with the frequency in the population of 1% or greater) and rare DNA mutations (less than 1% frequency) can generally be classified as genetic markers associated with endurance, power and strength (or combined power/strength) athlete status. It should be noted that other possible athlete statuses involving coordination and flexibility have still not been studied. The significance of a particular sport-related genetic marker is based on several criteria, such as type of the

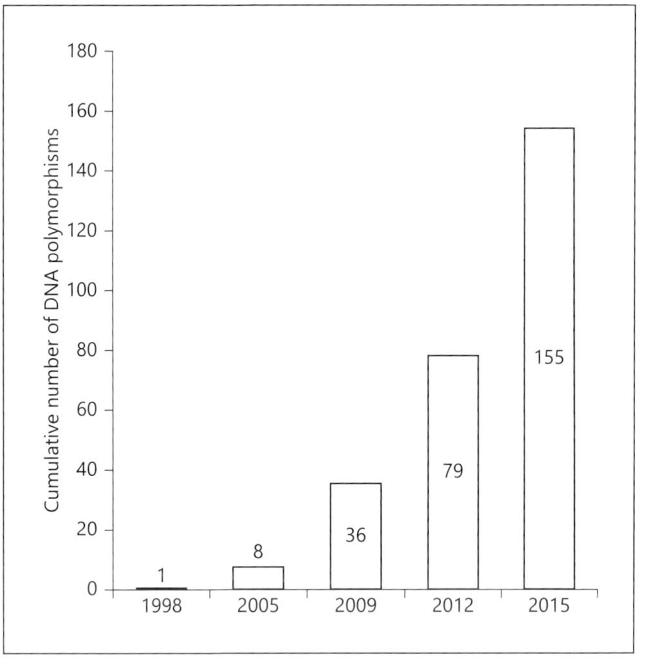

Fig. 1. Growth in the number of sports-related DNA polymorphisms discovered from 1998 to 2015.

polymorphism (stop loss/gain, frameshift, missense, synonymous, 3′/5′-UTR, intronic, non-coding RNA, 5′/3′-near gene, intergenic, etc.), its frequency in a given population, number of case-control and cross-sectional studies with positive or negative (controversial) results, total number of studied athletes, and supporting evidence from the functional studies (overexpression or knockout models, analysis of the luciferase activity with specific allele, etc.) [15].

Despite the obvious role of genetics in human athletic performance, there is little unequivocal evidence in support of a specific genetic variant with a major gene effect on a relevant performance phenotype, at least across the normal range of human trait distributions. This may be because complex traits are fundamentally polygenic (numerous genes with small effects), or because researchers failed to take into consideration the full range of environmental effects, or both. It is very important to note that each DNA locus can probably explain a very small proportion of the phenotypic variance (e.g. ~0.1 to ~1%) [4]. Therefore, very large sample sizes are needed to detect associations, and various combinatorial approaches should be used.

To date, several studies have sought to define or quantify the impact of multiple genotype combinations that influence human physical performance [16–34]. Figure 1 presents the cumulative number of sports-related DNA polymorphisms discovered from 1998 to 2015. By the end of June 2015, the total number of DNA polymorphisms with regard to sports genomics was 155. As the figure shows, most of these polymorphisms (77%) were discovered in the last 6 years (2010–2015), indicating a growing interest in the field of sports genomics and progress in DNA technologies.

The search for relevant publications was primarily based on the journals indexed in PubMed, SPORTDiscus and Google Scholar using a combination of key words (e.g. athletes, sport, exercise, physical performance, endurance, power, strength, training, gene, genetics, genotype, polymorphism, mutation). However, not all data from the articles were included in the current review due to language limitations, i.e. there were many more papers published in Chinese, German, Lithuanian, Russian, Spanish, Ukrainian, and other languages. Furthermore, data from the articles describing performance-associated polymorphisms investigated in the non-athletic cohorts or articles with very small cohorts of athletes (less than 25) and controls, as well as papers with mixed groups of athletes without stratification (e.g. when endurance athletes and sprinters were analysed as a combined group) were excluded from the current review.

A literature search revealed that at least 155 genetic markers (located within 82 autosomal genes, mitochondrial DNA, X and Y chromosomes) are linked to elite athlete status. These include 93 endurance-related genetic markers and 62 power/strength-related genetic markers (tables 1, 2). Importantly, 41 markers were identified within the last 2 years after performing GWASs of African-American, Jamaican, Japanese and Russian athletes, indicating that GWASs represent a promising and productive way to study sports-related phenotypes. Of note, 31 genetic markers [endurance markers: *ACE* I, *ACTN3* 577X, *ADRB2* 16Arg, *AQP1* rs1049305 C, *AMPD1* Gln12, *BDKRB2* –9, *COL5A1* rs12722 T, *GABPB1* rs12594956 A and rs7181866 G, *HFE* 63Asp, *KCNJ11* Glu23, mtDNA H haplogroup, mtDNA K haplogroup (unfavourable), *PPARA* rs4253778 G, *PPARD* rs2016520 C, *PPARGC1A* Gly482, *UCP3* rs1800849 T; power/strength markers: *ACE* D, *ACTN3* Arg577, *AGT* 235Thr, *AMPD1* Gln12, *CKM* rs1803285 G, *CREM* rs1531550 A, *GALNT13* rs10196189 G, *HIF1A* 582Ser, *IL6* rs1800795 G, *MTHFR* rs1801131 C, *NOS3* rs2070744 T, *PPARA* rs4253778 C, *PPARG* 12Ala, *SOD2* Ala16] have shown positive associations with athlete status in at least 2 studies and 12 of them (endurance markers: *ACE* I, *ACTN3* 577X, *HFE* 63Asp, *PPARA* rs4253778 G, *PPARGC1A* Gly482; power/strength markers: *ACE* D, *ACTN3* Arg577, *AMPD1* Gln12, *HIF1A* 582Ser, *MTHFR* rs1801131 C, *NOS3* rs2070744 T, *PPARG* 12Ala) in 3 or more studies. Conversely, the significance of 29 markers was not replicated in at least 1 study [for details, see 4], raising the possibility that several findings might be false-positive and require additional studies. Interestingly, almost all chromosomes (except for chromosome 20) include sport-related genetic markers.

According to existing data, endurance athlete status remains the most studied trait. Due to space limitations, there was no possibility to describe all 155 DNA polymorphisms in the current paper, but it should be noted that 120 of these genetic markers (mainly identified by candidate gene approach) were comprehensively characterized in a very recent review [4]. Given that, in this review we focus on the description of novel DNA polymorphisms recently identified by the GWAS approach.

Table 1. Gene variants (genetic markers) for endurance athlete status

Gene	Location	Polymorphism	Endurance-related marker
ACE	17q23.3	Alu I/D (rs4646994)	I
ACOXL	2q13	rs13027870 A/G	rs13027870 G
ACTN3	11q13.1	R577X (rs1815739 C/T)	577X
ADRA2A	10q24–q26	6.7/6.3 kb	6.7 kb
ADRB1	10q25.3	Ser49Gly (rs1801252 A/G)	49Gly
ADRB2	5q31–q32	Gly16Arg (rs1042713 G/A)	16Arg
ADRB3	8p12–8p11.1	Trp64Arg (rs4994 T/C)	64Arg
AGTR2	Xq22–q23	rs11091046 A/C	rs11091046 C
AQP1	7p14	rs1049305 C/G	rs1049305 C
AMPD1	1p13	Gln12X (rs17602729 C/T)	Gln12
BDKRB2	14q32.1–q32.2	+9/–9 (exon 1)	–9
		rs1799722 C/T	rs1799722 T
CAMK1D	10p13	rs11257754 A/G	rs11257754 A
CKM	19q13.32	rs8111989 A/G (*NcoI*)	rs8111989 A
CLSTN2	3q23	rs2194938 A/C	rs2194938 A
COL5A1	9q34.2–q34.3	rs12722 C/T (*BstUI*)	rs12722 T
		rs71746744 (AGGG/–)	rs71746744 AGGG
COL6A1	21q22.3	rs35796750 T/C	rs35796750 T
CPQ	8q22.2	rs6468527 A/G	rs6468527 A
EPAS1 (*HIF2A*)	2p21–p16	rs1867785 A/G	rs1867785 G
		rs11689011 C/T	rs11689011 T
GABPB1 (*NRF2*)	15q21.2	rs12594956 A/C	rs12594956 A
		rs8031031 C/T	rs8031031 T
		rs7181866 A/G	rs7181866 G
GALM	2p22.1	rs3821023 A/G	rs3821023 A
GNB3	2p13	rs5443 C/T (C825T)	rs5443 T
GRM3	7q21.1–q21.2	rs724225 A/G	rs724225 G
HFE	6p21.3	His63Asp (rs1799945 C/G)	63Asp
HIF1A	14q23.2	Pro582Ser (rs11549465 C/T)	Pro582
IGF1R	15q26.3	rs1464430 A/C	rs1464430 A
IL15RA	10p15.1	Asn146Thr (rs2228059 A/C)	146Thr
ITPR1	3p26.1	rs1038639 G/T	rs1038639 T
		rs2131458 C/T	rs2131458 T
FMNL2	2q23.3	rs12693407 A/G	rs12693407 G
KCNJ11	11p15.1	Glu23Lys (rs5219 C/T)	Glu23
L3MBTL4	18p11.31	rs17483463 C/T	rs17483463 T
MCT1 (SLC16A1)	1p12	Glu490Asp or A1470T (rs1049434 A/T)	Glu490
mtDNA loci	mtDNA	Haplogroups constructed from several mtDNA polymorphisms or single polymorphisms	G1
			H
			HV
			L0
			M*
			m.11215T, m.152C, m.15518T, m.15874G, m.4343G, $m.514(CA)_{\le 4}$, poly(C ≥7) stretch at m.568–573
			m.16080G
			m.5178C
			N9
			V
			Unfavourable: B
			Unfavourable: K
			Unfavourable: J2
			Unfavourable: T
			Unfavourable: L3*

Table 1. Continued

Gene	Location	Polymorphism	Endurance-related marker
NALCN-AS1	13q33.1	rs4772341 A/G	rs4772341 A
NATD1	17p11.2	rs732928 A/G	rs732928 G
NFATC4	14q11.2	Gly160Ala (rs2229309 G/C)	Gly160
NFIA-AS2	1p31.3	rs1572312 C/A	rs1572312 C
NOS3	7q36	Glu298Asp (rs1799983 G/T)	Glu298
		$(CA)_n$ repeats	164 bp
		27-bp repeats (4B/4A)	4B
		rs2070744 T/C (–786 T/C)	rs2070744 T
PPARA	22q13.31	rs4253778 G/C	rs4253778 G
PPARD	6p21.2–p21.1	rs2016520 T/C	rs2016520 C
		rs1053049 T/C	rs1053049 T
PPARGC1A	4p15.1	Gly482Ser (rs8192678 G/A)	Gly482
		rs4697425 A/G	rs4697425 A
PPARGC1B	5q32	Ala203Pro (rs7732671 G/C)	203Pro
		Arg292Ser (rs11959820 C/A)	292Ser
PPP3CA	4q24	rs3804358 C/G	rs3804358 C
PPP3CB	10q22.2	rs3763679 C/T	rs3763679 C
PPP3R1	2p15	Promoter 5I/5D	5I
RBFOX1	16p13.3	rs7191721 G/A	rs7191721 G
SGMS1	10q11.2	rs884880 A/C	rs884880 A
SLC2A4	17p13	rs5418 G/A	rs5418 A
SOD2	6q25.3	Ala16Val (rs4880 C/T)	C (Ala)
SPOCK1	5q31.2	rs1051854 G/T	rs1051854 T
TFAM	10q21	Ser12Thr (rs1937 G/C)	12Thr
TPK1	7q34–q35	rs10275875 C/T	rs10275875 T
TSHR	14q31	rs7144481 T/C	rs7144481 C
UCP2	11q13	Ala55Val (rs660339 C/T)	55Val
UCP3	11q13	rs1800849 C/T	rs1800849 T
VEGFA	6p12	rs2010963 G/C	rs2010963 C
VEGFR2	4q11–q12	His472Gln (rs1870377 T/A)	472Gln
Y chromosome haplogroups	Y chromosome	Haplogroups constructed from several Y chromosome polymorphisms	E*, E3* and K*(xP) Unfavourable: E3b1
ZNF429	19p12	rs1984771 A/G	rs1984771 G

Gene Variants and Endurance Athlete Status

A literature search revealed at least 93 markers are associated with endurance athlete status, and 19 of them were discovered by the use of microchip technology (table 1). Initially, Ahmetov et al. [7] examined the association between 1,140,419 SNPs and the relative maximal oxygen consumption rate ($\dot{V}O_{2\,max}$) in 80 international-level Russian endurance athletes (46 males and 34 females), and identified 6 suggestive 'endurance alleles' (with $p < 10^{-5}$ to 10^{-8}) which were replicated both in female and male subgroups. To validate the obtained results, the authors further performed case-control studies by comparing the frequencies of 6 SNPs between 218 endurance athletes (or 100 elite endurance athletes) and opposite cohorts (192 Russian controls, 1,367 European controls, and 230 Russian power athletes). It was assumed that the

Table 2. Gene variants (genetic markers) for power/strength athlete status

Gene	Location	Polymorphism	Power/strength-related marker
ACE	17q23.3	*AluI*/D (rs4646994)	D
ACTN3	11q13.1	R577X (rs1815739 C/T)	Arg577
ADRB2	5q31–q32	Gly16Arg (rs1042713 G/A)	Gly16
		Gln27Glu (rs1042714 C/G)	27Glu
AGT	1q42.2	Met235Thr (rs699 T/C)	235Thr
AGTR2	Xq22–q23	rs11091046 A/C	rs11091046 A
AMPD1	1p13	Gln12X (rs17602729 C/T)	Gln12
ARHGEF28	5q13.2	rs17664695 A/G	rs17664695 G
CACNG1	17q24	Gly196Ser (rs1799938 G/A)	196Ser
CALCR	7q21.3	rs17734766 A/G	rs17734766 G
CKM	19q13.32	rs8111989 A/G (*NcoI*)	rs8111989 G
CLSTN2	3q23	rs2194938 A/C	rs2194938 C
COTL1	16q24.1	rs7458 C/T	rs7458 T
CREM	10p11.21	rs1531550 G/A	rs1531550 A
DMD	Xp21.2	rs939787 C/T	rs939787 T
EPAS1 (HIF2A)	2p21–p16	rs1867785 A/G	rs1867785 G
		rs11689011 C/T	rs11689011 C
FOCAD	9p21	rs17759424 A/C	rs17759424 C
GABRR1	6q15	rs282114 A/G	rs282114 A
GALNT13	2q24.1	rs10196189 A/G	rs10196189 G
GPC5	13q32	rs852918 G/T	rs852918 T
HIF1A	14q21–q24	Pro582Ser (rs11549465 C/T)	582Ser
HSD17B14	19q13.33	rs7247312 A/G	rs7247312 G
IGF1	12q23.2	C–1245T (rs35767 C/T)	rs35767 T
IGF1R	15q26.3	rs1464430 A/C	rs1464430 C
IL1RN	2q14.2	VNTR 86-bp (intron 2)	*IL1RN*2*
IL6	7p21	–174 C/G (rs1800795 C/G)	rs1800795 G
IP6K3	6p21.31	rs6942022 C/T	rs6942022 C
MCT1 (SLC16A1)	1p12	Glu490Asp or A1470T (rs1049434 A/T)	490Asp
MED4	13q14.2	rs7337521 G/T	rs7337521 T
MPRIP	17p11.2	rs6502557 A/G	rs6502557 A
mtDNA loci	mtDNA	Haplogroups constructed from several mtDNA polymorphisms or single polymorphisms	F m.204C m.151T m.15314A Non-L/U6 Unfavourable: m.16278T, m.5601T, m.4833G, m.5108C, m.7600A, m.9377G, m.13563G, m.14200C, m.14569A
MTHFR	1p36.3	A1298C (rs1801131 A/C)	rs1801131 C
MTR	1q43	A2756G (rs1805087 A/G)	rs1805087 G
MTRR	5p15.31	A66G (rs1801394 A/G)	rs1801394 G
NOS3	7q36	rs2070744 T/C (–786 T/C)	rs2070744 T
		Glu298Asp (rs1799983 G/T)	Glu298
NRG1	8p12	rs17721043 A/G	rs17721043 A
PPARA	22q13.31	rs4253778 G/C	rs4253778 C
PPARG	3p25	Pro12Ala (rs1801282 C/G)	12Ala
PPARGC1B	5q32	rs10060424 C/T	rs10060424 C
RC3H1	1q25.1	rs767053 A/G	rs767053 G
SOD2	6q25.3	Ala16Val (rs4880 C/T)	C (Ala)
SUCLA2	13q14.2	rs10397 A/C	rs10397 A

Table 2. Continued

Gene	Location	Polymorphism	Power/strength-related marker
TPK1	7q34–q35	rs10275875 C/T	rs10275875 C
UCP2	11q13	Ala55Val (rs660339 C/T)	Ala55
VDR	12q13.11	*FokI* f/F (rs10735810 T/C)	rs10735810 T
WAPAL	10q23.2	rs4934207 C/T	rs4934207 C
ZNF423	16q12	rs11865138 C/T	rs11865138 C

'endurance allele' should be underrepresented in at least 1 opposite cohort (Russian controls or Russian power athletes) when compared to endurance athletes. This approach resulted in the remaining 3 SNPs (*NFIA-AS2* rs1572312 C, *TSHR* rs7144481 C, *RBFOX1* rs7191721 G) to be associated with endurance athlete status.

The *NFIA-AS2* gene encodes long non-coding RNA with undescribed function. Genes of antisense long non-coding RNAs are transcribed from either the same genomic site or a site distant from the gene locus where the sense transcript counterpart is produced. Antisense long non-coding RNAs repress – and in some cases can also activate – transcription of the targeted protein coding genes via mechanisms such as DNA methylation and chromatin modification at the genomic loci of the targeted genes. It was hypothesized that NFIA-AS2 is involved in the regulation of expression of the nuclear factor IA (*NFIA*) gene or erythroid/myeloid-specific RNAs. NFIA, as a transcription factor, induces erythropoiesis, whereas its silencing drives granulopoiesis. Consistent with the hypothesis, the authors also reported that the C allele was associated with activation of erythropoiesis (high level of haemoglobin, high number of reticulocytes and erythrocytes), while the A allele was associated with activated granulopoiesis (high number of neutrophils and greater leucocyte/erythrocyte ratio) [7].

As to the other 2 gene polymorphisms, it was established that RNA binding protein, fox-1 homolog *(Caenorhabditis elegans)* 1 (encoded by the *RBFOX1* gene) is an important splicing factor regulating developmental and tissue-specific alternative splicing in heart, muscle, and neuronal tissues [35]. Therefore, RBFOX1 is implicated in multiple medical conditions, including muscular dystrophies, cancers, neurodevelopmental and neuropsychiatric disorders. The thyroid-stimulating hormone receptor encoded by the *TSHR* gene is a membrane receptor for thyrotropin (produces thyroid hormones) and thyrostimulin (activates TSHR protein), and therefore a major controller of thyroid cell metabolism. Thyroid hormones are known as determinants of the metabolic and contractile phenotype of skeletal muscle [36]. TSHR also mediates the effect of thyrotropin on angiogenesis via cAMP-mammalian target of rapamycin signalling [37]. The rs7144481 polymorphism is located in the regulatory region (3′-UTR) of the *TSHR* gene.

With respect to the same groups of athletes and GWAS data using different criteria, such as (i) the SNP should be independently associated with $\dot{V}O_{2\,max}$ in male and female athletes separately (with $p < 10^{-3}$ adjusted for sex), (ii) the frequency of the endurance-related allele should be overrepresented in endurance athletes in comparison with controls and (iii) power athletes, the same group of authors identified an additional 6 alleles (*ZNF429* rs1984771 G, *FMNL2* rs12693407 G, *ACOXL* rs13027870 G, *ITPR1* rs2131458 A, *GALM* rs3821023 A, *NATD1* rs732928 G) with a suggested significance for the determination of endurance performance [38]. These SNPs are located in the genes involved in the regulation of lipid *(ACOXL)* and carbohydrate *(GALM)* metabolism, morphogenesis and cytokinesis *(FMNL2)*, intracellular Ca^{2+} signalling *(ITPR1)* and other processes *(ZNF429, NATD1)*.

In the third study of the same group, Galeeva et al. [39] performed a GWAS in 4 subgroups of Russian endurance athletes (n = 223; all and elite long-distance athletes, all and elite middle-distance athletes) and controls (n = 173), and found 93 SNPs associated with endurance athlete status with replications in all subgroups ($p < 10^{-4}$), but none of them reached a genome-wide significance level. Adding 3 criteria – (i) an increase in the frequency of the effect allele with an increase in the level of achievement of endurance athletes, (ii) significant differences in allelic frequencies between 56 elite endurance athletes and 67 elite power athletes (second case-control study), and (iii) a positive correlation of the effect allele with high values of $\dot{V}O_{2\,max}$ – resulted in the remaining 5 SNPs (effect alleles: *CAMK1D* rs11257754 A, *CPQ* rs6468527 A, *GRM3* rs724225 G, *SGMS1* rs884880 A, *L3MBTL4* rs17483463 T) to be associated with elite endurance athlete status. These SNPs are located in the genes involved in the regulation of carbohydrate metabolism *(CAMK1D)*, synthesis of thyroxine *(CPQ)*, glutamatergic neurotransmission *(GRM3)*, sphingomyelin and diacylglycerol metabolism *(SGMS1)* and chromatin modification *(L3MBTL4)*.

Finally, Gabdrakhmanova et al. [40] studied the differences in genomic profiles between Russian endurance and power athletes using the GWAS approach. At the first stage, by comparing genetic profiles of 2 groups of elite athletes (171 elite power and 56 elite endurance athletes), the authors identified 13 SNPs with suggestive significance (p values from 10^{-5} to 10^{-6}). At the second stage, they compared allelic frequencies of the discovered SNPs between 223 endurance athletes and 173 controls. As a final point, the regression analysis was performed to reveal the association with $VO_{2\,max}$ of endurance athletes (n = 71). These analyses resulted in the remaining 5 SNPs (*CLSTN2* rs2194938 A, *TPK1* rs10275875 T, *ITPR1* rs1038639 T, *NALCN-AS1* rs4772341 A, *SPOCK1* rs1051854 T) to be associated with endurance athlete status (based on case-control study and correlation with $VO_{2\,max}$). These SNPs are located in the genes involved in the regulation of neuronal excitability *(CLSTN2, NALCN-AS1)*, vitamin B_1 metabolism *(TPK1)*, muscle contraction *(ITPR1)*, and protein metabolism *(SPOCK1)*.

Gene Variants and Power/Strength Athlete Status

A literature search revealed at least 62 markers are associated with power/strength athlete status, and 22 of them were discovered by the use of microchip technology (table 2). By performing 3 GWASs of elite Jamaican (n = 95), African-American (n =

108), and Japanese (n = 54) sprint athletes and their matched controls (total number of controls: 617) and a subsequent meta-analysis, Wang et al. [41] found that the *CREM* A allele of the rs1531550 (G/A) polymorphism and the *GALNT13* G allele of the rs10196189 (A/G) polymorphism were significantly ($p < 2 \times 10^{-6}$) overrepresented in elite sprinters compared with controls. These results were also replicated ($p < 0.001$) in the Russian cohorts of sprinters and power/strength athletes [unpubl. data]. The *CREM* gene encodes a cAMP-responsive element modulator which is a bZIP transcription factor that binds to the cAMP-responsive element found in many viral and cellular promoters. It is an important component of cAMP-mediated signal transduction during the spermatogenetic cycle, as well as other complex processes [42]. *CREM* is highly expressed in the testis, heart, brain, pancreas, and retina. The UDP-N-acetyl-α-D-galactosamine:polypeptide N-acetylgalactosaminyltransferase 13 protein (encoded by *GALNT13*) is a member of the GALNT family, which initiates O-linked glycosylation of mucins by the initial transfer of N-acetylgalactosamine with an α-linkage to a serine or threonine residue and thus catalyses the initial reaction in O-linked oligosaccharide biosynthesis. *GALNT13* is highly expressed in the brain, B cells, kidney and liver and may be involved in metabolism and energy pathways [43].

In the GWAS of 483 Russian athletes (49 strength athletes, 103 endurance athletes and 331 athletes from other sports with a strength component: 89 sprinters, 38 strength/speed athletes, 64 wrestlers, 42 rugby players, 98 rowers/kayakers/canoers) and 173 controls, Egorova et al. [44] first identified 43 SNPs associated ($p < 10^{-5}$) with elite strength athlete status (when compared with controls), but none of them reached a genome-wide significance level. Adding 3 criteria – (i) an increase in the frequency of the effect allele with an increase in the level of achievement of strength athletes, (ii) significant differences in allelic frequencies between strength and endurance athletes, and (iii) at least 1 replication of association between effect alleles and predisposition to other sports with strength component – resulted in the remaining 8 SNPs [*SUCLA2* rs10397 A, *MED4* rs7337521 T, *GPC5* rs852918 T, *GABRR1* rs282114 A, *CACNG1* rs1799938 A (196Ser), *ARHGEF28* rs17664695 G, *WAPAL* rs4934207 C, *MPRIP* rs6502557 A alleles] with p values from 9.1×10^{-5} to 3.1×10^{-6}. These SNPs are located in the genes involved in the regulation of ATP production (*SUCLA2*), transcription of DNA (*MED4*), cell division and growth *(GPC5, ARHGEF28)*, neurotransmission *(GABRR1)*, muscle contraction *(CACNG1, MPRIP)*, and DNA repair *(WAPAL)*.

In another study, Ischenko et al. [45] performed a GWAS in 176 Russian power (89 sprinters, 38 speed/strength athletes and 49 strength athletes) athletes, a group of athletes with a speed/strength component (n = 204; 64 wrestlers, 42 rugby players, 98 rowers/kayakers/canoers), 223 endurance athletes and 173 controls. Initially, they performed 7 analyses using the GWAS data (elite power athletes vs. controls, all sprinters vs. controls, elite sprinters vs. controls, all speed/strength athletes vs. controls, elite speed/strength athletes vs. controls, all strength athletes vs. controls, elite strength athletes vs. controls) and found 68 SNPs which were associated with power athlete status (with p values from 0.001 to 1.34×10^{-5}) and replicated in all 3 subgroups of power athletes (regardless of their level of achievement). The comparison of allelic frequencies of these SNPs between the large cohort of power athletes (n = 380; i.e. power athletes plus group of athletes with speed/strength component) and endurance athletes (as a second control group) resulted in the remaining 8 SNPs (*PPARGC1B* rs10060424 C, *NRG1* rs17721043 A, *ZNF423* rs11865138 C, *RC3H1* rs767053 G, *IP6K3* rs6942022 C, *HSD17B14* rs7247312 G, *CALCR* rs17734766 G, *COTL1* rs7458 T) associated with power athlete status. These SNPs are located in the genes involved in the regulation of muscle fibre composition and carbohydrate/lipid metabolism *(PPARGC1B)*, growth and development *(NRG1, ZNF423)*, mRNA deadenylation and degradation *(RC3H1)*, metabolism of inositol hexakisphosphate *(IP6K3)*, metabolism of steroids *(HSD17B14)*, calcium homeostasis *(CALCR)* and actin cytoskeleton *(COTL1)*.

The comparison of genetic profiles of 492 elite Russian power/strength and 227 endurance athletes and controls revealed that the rare *DMD* rs939787 T allele was overrepresented in power/strength athletes (25.0%) compared to endurance athletes (8.8%; $p = 3.9 \times 10^{-9}$) and controls (16.3%; $p = 0.0354$). These results indicate that the *DMD* rs939787 T allele is favourable for power/strength performance [46]. The dystrophin (*DMD*) gene is the largest gene found in nature (2.4 Mb). Dystrophin is a large, rod-like cytoskeletal protein which is found at the inner surface of muscle fibres. Dystrophin is part of the dystrophin-glycoprotein complex, which bridges the inner cytoskeleton (F actin) and the extracellular matrix. In the subsequent study, by comparing genetic profiles of 2 groups of elite Russian athletes (171 elite power and 56 elite endurance athletes), and then between power athletes and controls, Gabdrakhmanova et al. [40] identified 3 SNPs (effect alleles: *CLSTN2* rs2194938 C, *FOCAD* rs17759424 C, *TPK1* rs10275875 C) associated with power athlete status. These SNPs are located in the genes involved in the regulation of neuronal excitability *(CLSTN2)*, cell growth *(FOCAD)*, and vitamin B_1 metabolism *(TPK1)*.

Conclusion

The current review provides evidence that at least 155 genetic markers are linked to elite athlete status. However, it should be emphasized that most (80%) of the case-control and association studies have not yet been replicated in independent samples. Based on that, we strongly believe that much more research is needed before these findings can be extended to practice in sport. On the other hand, since sport-related DNA polymorphisms do not fully explain the heritability of athlete status, other forms of variation, such as rare mutations and epigenetic markers (i.e. stable and heritable changes in gene expression), must be considered. The issues with respect to appropriate study designs, sample size, population stratification, and quality of the genotype/phenotype measurement are also of great importance. Future research should also be focused on identifying genetic markers associated with other sport-related phenotypes, such as flexibility, coordination and temperament of elite athletes.

The impact of genetics in sports and exercise appears to have multiple influences. Its positive effect on exercise performance must be combined with effective training programmes and favourable lifestyle habits for success in sports and health benefits [47]. Accordingly, one of the applications of sports genetics could be the development of predictive genetic performance tests, although it is still too premature currently in sports genomics to be able to definitively test for predictive genetic markers [48]. Furthermore, the application of genetic testing in sports could provide new opportunities for sports clubs to understand the athletes' susceptibility to certain pathological states (injuries, cardiomyopathies, sudden death, etc.), to map genetic suitability for specific team positions and roles, and to gain insights into the athletes' development in various sports or physical activities. Future research including multicentre GWASs and whole-genome sequencing in large cohorts of athletes with further validation and replication will substantially contribute to the discovery of large numbers of the causal genetic variants (mutations and DNA polymorphisms) that would partly explain the heritability of athlete status and related phenotypes.

References

1 De Moor MHM, Spector TD, Cherkas LF, Falchi M, Hottenga JJ, Boomsma DI, De Geus EJC: Genome-wide linkage scan for athlete status in 700 British female DZ twin pairs. Twin Res Hum Genet 2007;10:812–820.

2 Simoneau J-A, Bouchard C: Genetic determinism of fiber type proportion in human skeletal muscle. FASEB J 1995;9:1091–1095.

3 Alonso L, Souza E, Oliveira M, do Nascimento L, Dantas P: Heritability of aerobic power of individuals in northeast Brazil. Biol Sport 2014;31:267–270.

4 Ahmetov II, Fedotovskaya ON: Current progress in sports genomics. Adv Clin Chem 2015;70:247–314.

5 Eynon N, Nasibulina ES, Banting LK, Cieszczyk P, Maciejewska-Karlowska A, Sawczuk M, Bondareva EA, Shagimardanova RR, Raz M, Sharon Y, Williams AG, Ahmetov II, Lucia A, Birk R: The *FTO* A/T polymorphism and elite athletic performance: a study involving three groups of European athletes. PLoS One 2013;8:e60570.

6 Wang G, Mikami E, Chiu LL, de Perini A, Deason M, Fuku N, Miyachi M, Kaneoka K, Murakami H, Tanaka M, Hsieh LL, Hsieh SS, Caporossi D, Pigozzi F, Hilley A, Lee R, Galloway SD, Gulbin J, Rogozkin VA, Ahmetov II, Yang N, North KN, Ploutarhos S, Montgomery HE, Bailey ME, Pitsiladis YP: Association analysis of *ACE* and *ACTN3* in elite Caucasian and East Asian swimmers. Med Sci Sports Exerc 2013;45:892–900.

7 Ahmetov II, Kulemin NA, Popov DV, Naumov VA, Akimov EB, Bravy YR, Egorova ES, Galeeva AA, Generozov EV, Kostryukova ES, Larin AK, Mustafina LJ, Ospanova EA, Pavlenko AV, Starnes LM, Żmijewski P, Alexeev DG, Vinogradova OL, Govorun VM: Genome-wide association study identifies three novel genetic markers associated with elite endurance performance. Biol Sport 2015;32:3–9.

8 O'Connell K, Posthumus M, Collins M: *COL6A1* gene and Ironman triathlon performance. Int J Sports Med 2011;32:896–901.

9 Brown JC, Miller CJ, Posthumus M, Schwellnus MP, Collins M: The *COL5A1* gene, ultra-marathon running performance, and range of motion. Int J Sports Physiol Perform 2011;6:485–496.

10 Drozdovska SB, Dosenko VE, Ahmetov II, Ilyin VN: The association of gene polymorphisms with athlete status in Ukrainians. Biol Sport 2013;30:163–167.

11 Ahmetov II, Naumov VA, Donnikov AE, Maciejewska-Karłowska A, Kostryukova ES, Larin AK, Maykova EV, Alexeev DG, Fedotovskaya ON, Generozov EV, Jastrzębski Z, Zmijewski P, Kravtsova OA, Kulemin NA, Leonska-Duniec A, Martykanova DS, Ospanova EA, Pavlenko AV, Podol'skaya AA, Sawczuk M, Alimova FK, Trofimov DY, Govorun VM, Cieszczyk P: *SOD2* gene polymorphism and muscle damage markers in elite athletes. Free Radic Res 2014;48:948–955.

12 Ahmetov II, Hakimullina AM, Popov DV, Lyubaeva EV, Missina SS, Vinogradova OL, Williams AG, Rogozkin VA: Association of the *VEGFR2* gene His472Gln polymorphism with endurance-related phenotypes. Eur J Appl Physiol 2009;107:95–103.

13 Mustafina LJ, Naumov VA, Cieszczyk P, Popov DV, Lyubaeva EV, Kostryukova ES, Fedotovskaya ON, Druzhevskaya AM, Astratenkova IV, Glotov AS, Alexeev DG, Mustafina MM, Egorova ES, Maciejewska-Karłowska A, Larin AK, Generozov EV, Nurullin RE, Jastrzębski Z, Kulemin NA, Ospanova EA, Pavlenko AV, Sawczuk M, Akimov EB, Danilushkina AA, Zmijewski P, Vinogradova OL, Govorun VM, Ahmetov II: *AGTR2* gene polymorphism is associated with muscle fibre composition, athletic status and aerobic performance. Exp Physiol 2014;99:1042–1052.

14 Wang G, Padmanabhan S, Wolfarth B, Fuku N, Lucia A, Ahmetov II, Cieszczyk P, Collins M, Eynon N, Klissouras V, Williams A, Pitsiladis Y: Genomics of elite sporting performance: what little we know and necessary advances. Adv Genet 2013;84:123–149.

15 Ahmetov II, Donnikov AE, Trofimov DY: *ACTN3* genotype is associated with testosterone levels of athletes. Biol Sport 2014;31:105–108.

16 Williams AG, Dhamrait SS, Wootton PT, Day SH, Hawe E, Payne JR, Myerson SG, World M, Budgett R, Humphries SE, Montgomery HE: Bradykinin receptor gene variant and human physical performance. J Appl Physiol 2004;96:938–942.

17 Saunders CJ, Xenophontos SL, Cariolou MA, Anastassiades LC, Noakes TD, Collins M: The bradykinin b2 receptor (*BDKRB2*) and endothelial nitric oxide synthase 3 (*NOS3*) genes and endurance performance during Ironman triathlons. Hum Mol Genet 2006;15:979–987.

18 Akhmetov II, Popov DV, Mozhayskaya IA, Missina SS, Astratenkova IV, Vinogradova OL, Rogozkin VA: Association of regulatory genes polymorphisms with aerobic and anaerobic performance of athletes. Ross Fiziol Zh Im I M Sechenova 2007;93:837–843.

19 Ahmetov II, Popov DV, Astratenkova IV, Druzhevskaia AM, Missina SS, Vinogradova OL, Rogozkin VA: The use of molecular genetic methods for prognosis of aerobic and anaerobic performance in athletes. Hum Physiol 2008;34:338–342.

20 Williams AG, Folland JP: Similarity of polygenic profiles limits the potential for elite human physical performance. J Physiol 2008;586:113–121.

21 Eynon N, Meckel Y, Alves AJ, Yamin C, Sagiv M, Goldhammer E, Sagiv M: Is there an interaction between *PPARD* T294C and *PPARGC1A* Gly482Ser polymorphisms and human endurance performance? Exp Physiol 2009;94:1147–1152.

22 Gómez-Gallego F, Santiago C, González-Freire M, Muniesa CA, Fernández del Valle M, Pérez M, Foster C, Lucia A: Endurance performance: genes or gene combinations? Int J Sports Med 2009;30:66–72.

23 Ruiz JR, Gómez-Gallego F, Santiago C, González-Freire M, Verde Z, Foster C, Lucia A: Is there an optimum endurance polygenic profile? J Physiol 2009; 587:1527–1534.

24 Ruiz JR, Arteta D, Buxens A, Artieda M, Gómez-Gallego F, Santiago C, Yvert T, Morán M, Lucia A: Can we identify a power-oriented polygenic profile? J Appl Physiol 2010;108:561–566.

25 Muniesa CA, González-Freire M, Santiago C, Lao JI, Buxens A, Rubio JC, Martín MA, Arenas J, Gomez-Gallego F, Lucia A: World-class performance in lightweight rowing: is it genetically influenced? A comparison with cyclists, runners and non-athletes. Br J Sports Med 2010;44:898–901.

26 Santiago C, Ruiz JR, Muniesa CA, González-Freire M, Gómez-Gallego F, Lucia A: Does the polygenic profile determine the potential for becoming a world-class athlete? Insights from the sport of rowing. Scand J Med Sci Sports 2010;20:188–194.

27 Buxens A, Ruiz JR, Arteta D, Artieda M, Santiago C, González-Freire M, Martínez A, Tejedor D, Lao JI, Gómez-Gallego F, Lucia A: Can we predict top-level sports performance in power vs endurance events? A genetic approach. Scand J Med Sci Sports 2011;21: 570–579.

28 Eynon N, Ruiz JR, Meckel Y, Morán M, Lucia A: Mitochondrial biogenesis related endurance genotype score and sports performance in athletes. Mitochondrion 2011;11:64–69.

29 Hughes DC, Day SH, Ahmetov II, Williams AG: Genetics of muscle strength and power: polygenic profile similarity limits skeletal muscle performance. J Sports Sci 2011;29:1425–1434.

30 Ahmetov II, Gavrilov DN, Astratenkova IV, Druzhevskaya AM, Malinin AV, Romanova EE, Rogozkin VA: The association of *ACE*, *ACTN3* and *PPARA* gene variants with strength phenotypes in middle school-age children. J Physiol Sci 2013;63:79–85.

31 Ben-Zaken S, Meckel Y, Lidor R, Nemet D, Eliakim A: Genetic profiles and prediction of the success of young athletes' transition from middle- to long-distance runs: an exploratory study. Pediatr Exerc Sci 2013;25:435–447.

32 Egorova ES, Borisova AV, Mustafina LJ, Arkhipova AA, Gabbasov RT, Druzhevskaya AM, Astratenkova IV, Ahmetov II: The polygenic profile of Russian football players. J Sports Sci 2014;32:1286–1293.

33 Gineviciene V, Jakaitiene A, Tubelis L, Kucinskas V: Variation in the *ACE*, *PPARGC1A* and *PPARA* genes in Lithuanian football players. Eur J Sport Sci 2014; 14:289–295.

34 Ben-Zaken S, Meckel Y, Nemet D, Eliakim A. Genetic score of power-speed and endurance track and field athletes. Scand J Med Sci Sports 2015;25:166–174.

35 Kuroyanagi H: Fox-1 family of RNA-binding proteins. Cell Mol Life Sci 2009;66:3895–3907.

36 Simonides WS, van Hardeveld C: Thyroid hormone as a determinant of metabolic and contractile phenotype of skeletal muscle. Thyroid 2008;18:205–216.

37 Balzan S, Del Carratore R, Nicolini G, Beffy P, Lubrano V, Forini F, Iervasi G: Proangiogenic effect of TSH in human microvascular endothelial cells through its membrane receptor. J Clin Endocrinol Metab 2012;97:1763–1770.

38 Ahmetov II, Ischenko DS, Kulemin NA, Popov DV, Galeeva AA, Kostryukova ES, Alexeev DG, Egorova ES, Gabdrakhmanova LJ, Larin AK, Generozov EV, Ospanova EA, Pavlenko AV, Akimov EB, Vinogradova OL, Govorun VM: Genome-wide association study reveals seven genetic markers associated with maximal oxygen consumption rate in elite Russian endurance athletes. Eur J Hum Genet 2015;23(suppl 1):470.

39 Galeeva AA, Ischenko DS, Kulemin NA, Popov DV, Kostryukova ES, Alexeev DG, Egorova ES, Gabdrakhmanova LJ, Larin AK, Generozov EV, Ospanova EA, Pavlenko AV, Akimov EB, Vinogradova OL, Govorun VM, Ahmetov II: A multi-stage genome-wide association study of elite endurance athlete status. Eur J Hum Genet 2015;23(suppl 1):472.

40 Gabdrakhmanova LJ, Ischenko DS, Kulemin NA, Popov DV, Galeeva AA, Kostryukova ES, Alexeev DG, Egorova ES, Larin AK, Generozov EV, Ospanova EA, Pavlenko AV, Akimov EB, Vinogradova OL, Govorun VM, Ahmetov II: The difference in genomic profiles between endurance and power athletes. Eur J Hum Genet 2015;23(suppl 1):471.

41 Wang G, Padmanabhan S, Miyamoto-Mikami E, Fuku N, Tanaka M, Miyachi M, Murakami H, Cheng YC, Mitchell BD, Austin KG, Pitsiladis YP: GWAS of elite Jamaican, African American and Japanese sprint athletes. Med Sci Sports Exerc 2014;46(suppl):596–598.

42 De Cesare D, Fimia GM, Sassone-Corsi P: CREM, a master-switch of the transcriptional cascade in male germ cells. J Endocrinol Invest 2000;23:592–596.

43 Zhang Y, Iwasaki H, Wang H, Kudo T, Kalka TB, Hennet T, Kubota T, Cheng L, Inaba N, Gotoh M, Togayachi A, Guo J, Hisatomi H, Nakajima K, Nishihara S, Nakamura M, Marth JD, Narimatsu H: Characterization of a new human UDP-N-acetyl-α-D-galactosamine:polypeptide-N-acetyl galactosaminyl-transferase, designated pp-GalNac-T13, that is specifically expressed in neurons and synthesizes GalNac α-serine/threonine antigen. J Biol Chem 2003;278:573–584.

44 Egorova ES, Ischenko DS, Kulemin NA, Galeeva AA, Kostryukova ES, Alexeev DG, Gabdrakhmanova LJ, Larin AK, Generozov EV, Ospanova EA, Pavlenko AV, Govorun VM, Ahmetov II: Genome-wide association study of elite strength athlete status in Russians. Eur J Hum Genet 2015;23(suppl 1):468–469.

45 Ischenko DS, Galeeva AA, Kulemin NA, Kostryukova ES, Alexeev DG, Egorova ES, Gabdrakhmanova LJ, Larin AK, Generozov EV, Ospanova EA, Pavlenko AV, Govorun VM, Ahmetov II: Genome-wide association study of elite power athlete status. Eur J Hum Genet 2015;23(suppl 1):472.

46 Naumov VA, Ahmetov II, Larin AK, Generozov EV, Kulemin NA, Ospanova EA, Pavlenko AV, Kostryukova ES, Alexeev DG, Govorun VM: Genome-wide association analysis identifies a locus on *DMD* (dystrophin) gene for power athlete status in Russians. Eur J Hum Genet 2014;22(suppl 1):502.

47 Pokrywka A, Kaliszewski P, Majorczyk E, Zembroń-Łacny A: Genes in sport and doping. Biol Sport 2013;30:155–161.

48 Webborn N, Williams A, McNamee M, Bouchard C, Pitsiladis Y, Ahmetov I, Ashley E, Byrne N, Camporesi S, Collins M, Dijkstra P, Eynon N, Fuku N, Garton FC, Hoppe N, Holm S, Kaye J, Klissouras V, Lucia A, Maase K, Moran C, North KN, Pigozzi F, Wang G: Direct-to-consumer genetic testing for predicting sports performance and talent identification. Br J Sports Med 2015;49:1486–1491.

Dr. Ildus I. Ahmetov
Sport Technology Research Center
Volga Region State Academy of Physical Culture, Sport and Tourism
35, Universiade Village, 420138 Kazan (Russia)
E-Mail genoterra@mail.ru

Posthumus M, Collins M (eds): Genetics and Sports, ed 2, revised, extended.
Med Sport Sci. Basel, Karger, 2016, vol 61, pp 55–67 (DOI: 10.1159/000445241)

The Future of Genomic Research in Athletic Performance and Adaptation to Training

Guan Wang[a] · Masashi Tanaka[d] · Nir Eynon[f] · Kathryn N. North[g] · Alun G. Williams[b] · Malcolm Collins[h] · Colin N. Moran[c] · Steven L. Britton[i] · Noriyuki Fuku[e] · Euan A. Ashley[j] · Vassilis Klissouras[k] · Alejandro Lucia[l] · Ildus I. Ahmetov[m] · Eco de Geus[n] · Mohammed Alsayrafi[o] · Yannis P. Pitsiladis[a]

[a]FIMS Reference Collaborating Centre of Sports Medicine for Anti-Doping Research, University of Brighton, Eastbourne, [b]Department of Exercise and Sport Science, Manchester Metropolitan University, Crewe, and [c]Health and Exercise Sciences Research Group, University of Stirling, Stirling, UK; [d]Department of Longevity and Health, Tokyo Metropolitan Institute of Gerontology, Tokyo, and [e]Graduate School of Health and Sports Science, Juntendo University, Chiba, Japan; [f]Institute of Sport, Exercise and Active Living (ISEAL), Victoria University, and [g]Murdoch Children's Research Institute, Royal Children's Hospital, Melbourne, Vic., Australia; [h]Division of Exercise Science and Sports Medicine, Department of Human Biology, Faculty of Health Sciences, University of Cape Town, Cape Town, South Africa; [i]Department of Anesthesiology, University of Michigan Medical School, Ann Arbor, Mich., and [j]Stanford University Medical Center, Stanford, Calif., USA; [k]Ergophysiology Research Laboratory, Department of Sport Medicine and Biology of Physical Activity, University of Athens, Athens, Greece; [l]School of Doctorate Studies and Research, Universidad Europea de Madrid, Madrid, Spain; [m]Sport Technology Research Center, Volga Region State Academy of Physical Culture, Sport and Tourism, Kazan, Russia; [n]VU University and VU Medical Centre, Amsterdam, The Netherlands; [o]Anti-Doping Lab Qatar (ADLQ), Doha, Qatar

Abstract

Despite numerous attempts to discover genetic variants associated with elite athletic performance, an individual's trainability and injury predisposition, there has been limited progress to date. Past reliance on candidate gene studies focusing predominantly on genotyping a limited number of genetic variants in small, often heterogeneous cohorts has not generated results of practical significance. Hypothesis-free genome-wide approaches will in the future provide more comprehensive coverage and in-depth understanding of the biology underlying sports-related traits and related genetic mechanisms. Large, collaborative projects with sound experimental designs (e.g. clearly defined phenotypes, considerations and controls for sources of variability, and necessary replications) are required to produce meaningful results, especially when a hypothesis-free approach is used. It remains to be determined whether the novel approaches under current implementation will result in findings with real practical significance. This review will briefly summarize current and future directions in exercise genetics and genomics.

World-class sporting performance is a polygenic, multifactorial trait, determined by the interaction of genes and the environment. This review will provide a brief summary of the current status in exercise genetics reflecting the fact that very little of practical significance is known and will focus on future directions in exercise genetics and genomics.

Overview of the Current Status

There have been numerous efforts over the past few decades to discover performance-related genes [1–12]. Commonly used gene discovery methods in genetic research have included family-based studies (e.g. twin studies and linkage analyses) and association studies in unrelated population samples (i.e. candidate gene and hypothesis-free approaches). There have been numerous attempts to quantify the heritable component of athletic performance from twins and family aggregation studies. For example, the heritability of athletic status, trainability, and exercise behaviour is estimated to be 66, 47, and 62%, respectively [13–15]. Linkage analyses have revealed a number of chromosomal regions that contain predisposing genes associated with performance-related phenotypes such as maximal oxygen uptake and the response to training [16–19], maximal power output [18, 19], cardiac output [20], physical activity levels [13, 21, 22], and muscle strength-related traits [23, 24]. Early family studies showed that genetic factors contribute to overall performance and performance-related traits. On the other hand, genetic association studies that are conducted in unrelated case-control samples and directly compare the allele frequencies of marker(s) in the candidate regions or spanning throughout the whole genome have a greater power in identifying variants of modest effect and of lower frequency. As of 2008, over 200 genes were reported to be significantly associated with human physical performance [6], and there have been many more associations reported since. To date, most of the genetic findings in athletic performance have been generated using the candidate gene association approach (in small, often heterogeneous cohorts), and the majority of associations have been inconclusive due primarily to (i) the variants genotyped are not causal and provide incomplete linkage with other functional variants, (ii) studies are underpowered, (iii) population stratification, and (iv) phenotypic and locus heterogeneity. False-positive discoveries can also occur in studies examining multiple genes or splitting the cohorts into subgroups for separate analysis [25] without controlling for multiple testing. Among all the genes reported, the angiotensin-1-converting enzyme insertion/deletion (*ACE* I/D) polymorphism and the α-actinin-3 *(ACTN3)* R577X polymorphism have been the most extensively studied and, in general, have consistently been associated with elite endurance (*ACE* I allele) and sprint performance (*ACTN3* R allele). The gene discovery methods, their drawbacks and advantages, and the outcome from using these approaches related to sports genetics are presented elsewhere in more detail [26–28].

Despite the inconsistent findings reported in the exercise genetics literature, this genetic knowledge has already been used by companies to produce sports-related testing kits that are offered to coaches, parents, and athletes. The first testing kit available on the market dated back to 2004 [29], not long after Yang et al. [30] had reported the potential influence of *ACTN3* on sports performance. A recent review examined the issues surrounding genetic testing for athletic ability, searched and summarized the number of commercially available testing kits [31]. The consensus view in the scientific community is that current genetic findings have no predictability in personalized training and talent identification; therefore, current genetic testing using this information is misleading and must not be used for the purposes indicated. The search conducted as part of that review showed that, among 39 identified companies providing genetic tests, 21 did not reveal the specific genetic variants under testing [31]. The remaining 18 companies presented the variant details [31]. The number of variants tested by the 18 remaining companies ranged from 1 to 27, with the top 2 commonly tested variants being *ACE* I/D (11 of 18) and *ACTN3* R577X (16 of 18) polymorphisms [31]. Genetic testing is further reviewed in the paper by Williams et al. [this vol., pp. 105–119] of this book.

Athletic performance is a complex trait that requires different genetically driven components of human biological systems to coordinate effectively with the environment so as to excel. It is undoubtedly the case that multiple genes are associated with the performance trait and/or its components, though current progress in understanding the genetic architecture underlying these traits is very limited. Exercise genomic science requires an urgent paradigm shift [32] involving a collection of different approaches to discover the genetic basis of human performance, talent selection, trainability, and injury prevention. Ethical issues around genetic testing are paramount particularly in young children and need to be considered carefully at all times [29, 31].

The Paradigm Shift to the Genomics Era

Previous candidate gene studies have focused on a limited number of variants in small and often heterogeneous cohorts resulting inevitably in spurious and conflicting results. It is necessary that exercise genomics are not restricted to the current and common practice of focusing on candidate genes, typically defined by the author preferences or from biases in the published literature, and the reliance on small, statistically underpowered, observational studies [32]. It is vital that exercise genomic science needs to shift to an unbiased exploration of the genome using all the power of genomics, epigenomics, and transcriptomics in combination with large observational (preferably prospective) and experimental study designs with the emphasis firmly on replication [32]. Therefore, large, collaborative efforts are necessary for meaningful progress to be made in the area of exercise genomics.

A group of main investigators in sports and exercise science gathered together in a symposium held in Santorini, on May 14–17, 2015, reviewed the main findings in exercise genetics and genomics, and explored and discussed future trends and possibilities. At the end of the symposium, the participants agreed to launch a large collaborative initiative (named 'the Athlome Project Consortium', www.athlomeconsortium.org) to share and bring together resources for future investigations. This unique initiative is envisaged to help bring about a real paradigm shift to large collaborative projects using all available state-of-the-art advances (i.e. 'omics' technologies) as recently described [33]. Briefly, the main goals of the Athlome Project are:

1 To establish an ethically sound international research consortium (Athlome Project Consortium) and biobank resource systematically across individual centres;

2 To discover genetic variants associated with exercise performance, adaptive response to exercise training, and skeletal-muscle injuries using the genome-wide association study (GWAS) approach, targeted sequencing, or whole-genome sequencing, where possible;

3 To validate and replicate the genetic markers from the discovery phase across sex and ethnicity;

4 To conduct functional investigations following replicated findings (e.g. study the replicated single-nucleotide polymorphisms, SNPs, and their linkage disequilibrium regions, in vitro expression studies and knockouts of nearby genes) to better understand the associated biology.

Understanding genetic variation underlying athletic performance, adaptation to exercise training (in both human and animal models), and exercise-related musculoskeletal injuries is the primary focus of the Athlome Project Consortium. In addition, other 'omics' approaches, such as epigenomics, transcriptomics, proteomics and metabolomics, will need to be adopted to help dissect the molecular mechanisms of elite athletic performance, response to training, and injury predisposition. Amongst the numerous challenges will be the application of effective statistical analyses to minimize confounding effects due to artefacts on the exceedingly large data sets originating from different research centres, populations, and analytical platforms.

Current investigators and research centres participating in the Athlome Project Consortium have been described elsewhere [33] and are illustrated below.

Eastern Europe Population Studies (The Russian and Belarusian Cohorts, GELAK, GELAV, and GUAP)

The Russian and Belarusian cohorts, the Genetics and Epigenetics of Lithuanian Athletes from Kaunas (GELAK) and Vilnius (GELAV), and the Genome of Ukrainian Athletes Project (GUAP) have consolidated to identify genetic and epigenetic variations associated with high-level sports performance. The cohort comprises East Europeans (from Belarus, Lithuania, Russia, and Ukraine; in total n = 8,228 athletes and n = 4,121 controls). The athletes are grouped into international (including participants in Olympics and World Championships), national, regional, or local/non-competitive categories. These include biathletes, distance runners, cyclists, triathletes, kayakers, rowers, canoers, modern pentathletes, orienteers, skiers, speed skaters, short-trackers, walkers, weightlifters, bodybuilders, powerlifters, strongmen, sprint runners (≤400 m), sprint swimmers (50–100 m), decathletes, heptathletes, combat athletes, field athletes, bobsleigh athletes, rhythmic and artistic gymnasts, figure skaters, fencers and team ball sport players. A portion of the participants have been evaluated with a variety of quantitative performance- and health-related assessments, including strength/power-related measurements, agility/speed-related measurements, balance, flexibility and coordination measurements, endurance-related measurement, skeletal muscle biopsy, and health-related measurements. Current analyses include whole-exome sequencing in 400 (100 athletes and 300 controls) and whole-genome genotyping (1.14 million SNPs) in 1,700 (1,400 athletes and 300 controls) subjects.

The principal investigators are: Ildus I. Ahmetov (Volga Region State Academy of Physical Culture, Sport and Tourism, Russia), Svitlana B. Drozdovska (National University of Physical Education of Ukraine, Ukraine), Colin N. Moran (University of Stirling, UK), Valentina Ginevičienė (Vilnius University, Lithuania), Andrei A. Gilep (Institute of Bioorganic Chemistry NASB, Belarus).

ELITE Study

The Exercise at the Limit – Inherited Traits of Endurance (ELITE) consortium is a global initiative with the main objective to map the role that genetics plays in athletic ability versus environmental factors, such as training. Study participant (n >500) selection is based on a physiological variable relevant for both health and sport performance, i.e. maximum oxygen uptake ($\dot{V}O_{2\,max}$). The main inclusion criterion is $\dot{V}O_{2\,max}$ >75 ml/kg/min for men and >63 ml/kg/min for women, respectively. The consortium is continuously expanding and is recruiting athletes from all over the globe (with the main focus on Caucasians, North East Africans, East Asians and South Americans) who are successful in endurance sports (running, cycling, cross-country skiing, triathlon, and rowing). Analyses currently include enhanced whole-exome sequencing and GWAS (1.7 million SNPs). The combination of analytical methods will enable findings and differentiation between common variants with small effects and novel rare variants with larger effects. The aim is also to investigate gender and ethnic differences.

The principal investigators are: Euan A. Ashley, C. Mikael Mattsson, Matthew Wheeler, Daryl Waggott (Stanford University, USA).

Elite East African Athlete Cohort

The consortium also aims to study the East African running success by analysing data from previously recruited subjects: (i) 76 endurance runners (64 men) and 38 sprint and power event athletes (18 men) from the Ethiopian national athletics teams, 315 controls from the general Ethiopian population (281 men), 93 controls from the Arsi region of Ethiopia (80 men), and (ii) 291 elite Kenyan endurance athletes (232 men) and 85 control participants (40 men). Seventy (59 men) Kenyan athletes had competed internationally and achieved outstanding success.

The principal investigators are: Yannis Pitsiladis (University of Brighton, UK), Robert Scott (University of Cambridge, UK).

GAMES Study

An international consortium (GAMES) was established to compare allele frequencies between elite endurance athletes and ethnicity-matched controls [34]. GWASs were undertaken on 2 cohorts of elite endurance athletes (GENATHLETE and Japanese endurance runners) and their respective controls, from which a panel of 45 candidate SNPs was identified. These markers were tested for replication in 7 additional cohorts of endurance athletes and controls from Australia, Ethiopia, Japan, Kenya, Poland, Russia, and Spain. The study is based on a total of 1,520 endurance athletes (835 of them had competed in World Championships or Olympic Games) and 2,760 controls.

The principal investigators are: Claude Bouchard, Tuomo Rankinen (Pennington Biomedical Research Centre, Louisiana State University, USA), Noriyuki Fuku (Juntendo University, Japan), Yannis Pitsiladis (University of Brighton, UK), Bernd Wolfarth (Humboldt University, Germany), Alejandro Lucia (Universidad Europea de Madrid, Spain).

GENATHLETE Study

The study was launched in 1993 with the aim of identifying DNA variants that are present at different frequencies between elite endurance athletes and sedentary controls. Male endurance athletes and controls were recruited from Canada, Finland, Germany, and the USA. The cohort assembled to date includes 315 elite endurance athletes and 320 matched controls. Selection criteria for the all-male endurance athlete sample include that they had to be athletes of national or international caliber with a $\dot{V}O_{2\,max}$ of at least 75 ml/kg/min. The mean value for the 315 athletes is currently 79 ml/kg/min while the mean for the 320 control subjects reached 40 ml/kg/min. Multiple candidate genes have been studied using the resources of GENATHLETE. A genome-wide screen for common variants has been performed on GENATHLETE (see GAMES cohort above), and further studies are focusing on nuclear and mitochondrial DNA sequencing.

The principal investigators are: Claude Bouchard, Tuomo Rankinen (Pennington Biomedical Research Centre, Louisiana State University, USA), Bernd Wolfarth (Department of Sports Medicine, Charité Medical School, Germany), Louis Pérusse (Laval University, Canada), Rainer Rauramaa (University of Eastern Finland, Finland).

GENESIS Study

The Genetics of Elite Status in Sport (GENESIS) consortium aims to identify molecular genetic characteristics associated with successful sports performance. The cohort (current n >1,200) is mainly composed of UK athletes. Sports include marathon running and other track-and-field athletics, cycling and team sports (e.g. soccer). The RugbyGene Study is a major subcomponent of GENESIS and focuses on rugby (both union and league codes). Objectives of GENESIS are: (i) to increase the current cohort size substantially; (ii) to apply hypothesis-free approaches to identify molecular genomic markers; (iii) to expand GENESIS from genomics to other 'omics', and (iv) to combine the 'omics' data with athlete health and performance data to maximize the practical impact of GENESIS.

The principal investigators are: Alun G. Williams, Stephen H. Day, Georgina K. Stebbings (Manchester Metropolitan University, UK), Robert M. Erskine (Liverpool John Moores University, UK), Hugh E. Montgomery (University College London, UK).

Gene SMART Study

The Gene SMART (Skeletal Muscle Adaptive Response to Training) study aims to identify the gene variants that predict the skeletal muscle response to both a single bout and 4 weeks of high-intensity interval training in 3 different training centres. While the lead training and testing centre is located in Victoria University, Melbourne, 2 other centres have been launched at Bond University, Gold Coast, Australia, and the University of São Paulo, Brazil. The cohort is comprised of moderately trained, healthy male participants (aged 20–45 years, body mass index ≤30). Participants are undergoing similar exercise testing and exercise training in 3 different laboratories. Dietary habits are assessed by questionnaire and nutritionist consultation. Activity history is assessed by questionnaire, and the current activity level is assessed by activity monitoring. A number of muscle and blood analyses are to be performed, including genotyping, mitochondrial respiration, transcriptomics, proteomics, and enzyme activity before, during and after training, where appropriate. Currently ~40 participants have finished the study, and the aim is to train a total of 250 participants. The Gene SMART study also includes baseline and posttraining testing and sampling for all participants.

The principal investigators are: David Bishop, Nir Eynon (Victoria University, Australia).

GOINg Study

The recently established Genomics of Injuries (GOINg) consortium aims to identify DNA variants that modify the risk of anterior cruciate ligament injuries. It is the only consortium within the Athlome Project to specifically investigate exercise-associated musculoskeletal injuries. The plan is to screen current known loci for anterior cruciate ligament injury susceptibility in larger data sets in an attempt to determine whether they remain as susceptibility loci across all populations using the hypothesis-driven candidate gene case-control study design. Care will be taken to use the same criteria to accurately phenotype, with respect to ancestry, sporting and occupational details, injury profile and mechanism(s) of injury, other injury history and family history, as well as other appropriate medical history and medication use. The actual functional significance of the identified variants will also be investigated. This initial phase will be followed by sequencing, and the research objectives will be eventually expanded to include other 'omics'. Thus far, an anterior cruciate ligament rupture consortium has collected DNA samples and clinical as well as physical and occupational activity information from subjects from South Africa, Poland, Australia, Russia and Italy.

The principal investigators are: Malcolm Collins, Alison September, Michael Posthumus (University of Cape Town, South Africa), Nir Eynon (Victoria University, Australia), Pawel Cieszczyk (University of Szczecin, Poland).

J-HAP Study

The Japanese Human Athlome Project (J-HAP) focuses on the study of genes associated with physical performance and its related phenotypes (e.g. muscle mass, muscle fibre type, $\dot{V}O_{2\ max}$). The cohort is comprised of Japanese athletes (currently >2,400, mainly international and national levels) and healthy Japanese controls (currently >1,000). These athletes are mainly track-and-field athletes and swimmers competing in endurance- and sprint/power-oriented events. Multiple 'omics' approaches will be used to determine genes in talent identification in the Japanese population. Among the collected Japanese athlete and control samples, approximately 200 muscle biopsies were obtained from both athletes and controls in order to investigate genetic variants associated with muscle fibre type distribution. In addition, saliva samples from more than 1,200 athletes with injury history were collected to identify genetic variants associated with sports injuries.

The primary investigators are: Noriyuki Fuku (Juntendo University, Japan), Naoki Kikuchi (Nippon Sport Science University, Japan), Eri Miyamoto-Mikami (National Institute of Fitness and Sports in Kanoya, Japan).

NTR Study

The Netherlands Twin Register (NTR) is a population-based cohort recruiting both newborn and adult multiples and their family members with continuous longitudinal data collection. In the past 25+ years, around 40% of all twins and multiples in the Netherlands have taken part in the NTR research projects. Family members and spouses of twins also took part, leading to a total of over 185,000 participants across multiple research projects. The longitudinal information that has been collected extends from genotype to biomarkers, gene expression to rich behavioural information including biennial reports on (competitive) sports participation and performance level and on injuries related to sports. In its sports research track, the NTR aims to understand the interplay between genetic and environmental factors shaping individual differences in sports participation and performance. In the NTR, participants are recruited as newborns and followed into young adulthood; 520 have played competitively at a regional and 189 at a national level. Main sports that Dutch adolescents/young adults engage in are swimming, tennis, bicycling, soccer, and field hockey. The longitudinal data collection of the NTR is ongoing and securely funded for the next 5 years.

The principal investigators are: Eco de Geus, Meike Bartels (VU University and VU Medical Centre, the Netherlands).

POWERGENE Study

The POWERGENE consortium aims to characterize the elite sprint/power athlete genotype. The internationally competitive (Olympic/World Championship qualifiers) sprint/power athletes are from: Australia, Belgium, Greece, Italy, Jamaica, Japan, Lithuania, Poland, Spain, the USA, Brazil, and Russia. They will be compared with subelite athletes (national qualifiers), endurance athletes, team athletes and controls. The current cohort consists of female (n = 264) and male (n = 481) specialist power athletes across 3 major ethnicities (i.e. European, West African and East Asian ancestries). Sprint/power athletes include those individuals competing in track (≤800 m) and field (jump, throw) events, cycling (track), swimming (≤200 m), gymnastics (artistic), weightlifting, judo, speed skating and power lifting. Endurance athletes (n = 586) include track and road running specialists (>800 m), rowers, cyclists, swimmers (>200 m), triathletes, and ironmen. Team sports (n = 862) include football (soccer), cricket, hockey, volleyball, and basketball.

The principal investigators are: Yannis Pitsiladis (University of Brighton, UK), Kathryn North (Murdoch Children's Research Institute, Australia), Nir Eynon (Victoria University, Australia).

Super-Athletes: Genes and Sweat

The study aims to (i) identify genetic variants associated with elite athletic performance, (ii) study potential ethnic differences, and (iii) study the functional significance of the identified variants. A GWAS will be carried out in 3,000 consenting elite athletes, tested negative for doping substances at the Anti-Doping Laboratories, Federazione Medico Sportiva Italiana (FMSI) and Anti-Doping Lab Qatar (ADLQ), using Illumina genotyping technologies. Examining genotype frequency distribution of elite athletes from European countries (where most of the FMSI samples will be obtained) against those from South Asian and African countries (where most of the ADLQ samples are expected to be obtained) would help to identify potential ethnic differences in the genetic predisposition to athletic performance. Subsequently, the urine metabolome in a subset of these athletes (1,000 subjects) will be assessed and related to the athlete's sporting discipline.

The principal investigators are: Mohamed El-Rayess, Costas Georgakopoulos, Mohammed Alsayrafi (ADLQ, Qatar), Francesco Botre (FMSI, Italy), Karsten Suhre (Weill Cornell Medical College in Qatar, Qatar), Mike Hubank (University College London, UK).

Epigenetics of Elite Athletic Performance Study

It is clear from animal and human studies that epigenetic marks play a role in the modulation of gene expression in relevant tissues. There are also indications that epigenetic marks can be altered by acute and chronic exercise in skeletal muscle and adipose tissue where they have been studied. Thus, individual differences in any exercise-related traits can potentially be explained by not only the impact of DNA sequence variation on biology and behaviour, but also by the effects of epigenomic signalling on gene expression. We are formulating the hypothesis that elite athletic performance is influenced by epigenomic alterations, facilitating morphological, physiological, metabolic, cognitive, emotional and behavioural changes that empower the athlete to push performance beyond existing boundaries. We envisage testing this hypothesis by recruiting twin athletes competing at the Olympic or World Championship levels.

The principal investigators are: Vassilis Klissouras (University of Athens, Greece), Yannis Pitsiladis (University of Brighton, UK).

Rat Models of Exercise and Health

The purpose of the low-capacity rat/high-capacity rat model is to serve as a resource for the in-depth study of rat models to resolve the extremes of exercise and health. By connecting clinical observation with a theoretical base, the working hypothesis is that *variation in capacity for energy transfer is the central mechanistic determinant between disease and health (energy transfer hypothesis)*. As an unbiased test of this hypothesis, this study showed that two-way artificial selective breeding of rats for low and high intrinsic endurance exercise capacity also produced rats that differed for numerous disease risks, including the metabolic syndrome, premature ageing, fatty liver disease, obesity, and Alzheimer's disease. Exercise capacity is a result of intrinsic capacity plus adaptation to all aspects of physical activity. To capture this biology, rats for low and high response to 8 weeks of treadmill running exercise were selectively bred. Thus, the study has models that represent the 4 'corners' of exercise capacity. These contrasting animal model systems may prove to be translationally superior relative to more widely used simplistic models for understanding disease conditions. The rat models may be thoroughly explored to discover causal mechanisms and develop effective therapeutics. These rats are being studied at over 50 institutions in 11 countries.

The principal investigators are: Steven Britton, Lauren Koch (University of Michigan, USA).

1,000 Athlomes Project

The 1,000 Athlomes Project aims to sequence 1,000 genomes of sprinters and distance runners of West and East African descent. Phase 1 of the project has already commenced and involves the sequencing of 12 sprinters and 12 distance runners of the highest level (i.e. world record holders, Olympians and World Champions). Phase 2 (2016–2018) will involve increasing the sample size for sequencing to 100 genomes. The pool of the runners to be sequenced will be expanded to 1,000 by 2020 (phase 3). An important aim of this sequencing project is to document the genotype distribution of elite East and West African athletes. The large amount of genotype data to be generated from the 1,000 Athlome Project will serve (1) as a reference panel for future performance studies and (2) to guide other extreme phenotype studies in medical science.

The principal investigators are: Masashi Tanaka (Tokyo Metropolitan Institute of Gerontology, Japan), Yannis Pitsiladis (University of Brighton, UK).

Conclusion

The Athlome Project Consortium is a unique and highly ambitious attempt to oversee an essential paradigm shift in the area of exercise genomics. By presenting here the main study cohorts and projects that are currently included in the Athlome Consortium, it is our intention to show a global view of the main studies and initiatives that will be performed in the foreseeable future in the field of sports genomics and that are likely to provide exciting new findings of real practical significance. It is timely, therefore, that joint efforts are being made by the main investigators in sports and exercise genetics to uncover genetic variations underlying human physical performance. Although extraordinary challenges remain in the coming years, this unique collaborative initiative has the greatest chance to succeed where individual efforts have failed, to increase our understanding of the biology of exercise performance and related performance traits.

References

1 Rankinen T, Pérusse L, Rauramaa R, Rivera MA, Wolfarth B, Bouchard C: The human gene map for performance and health-related fitness phenotypes: the 2001 update. Med Sci Sports Exerc 2002;34: 1219–1233.

2 Pérusse L, Rankinen T, Rauramaa R, Rivera MA, Wolfarth B, Bouchard C: The human gene map for performance and health-related fitness phenotypes: the 2002 update. Med Sci Sports Exerc 2003;35: 1248–1264.

3 Rankinen T, Pérusse L, Rauramaa R, Rivera MA, Wolfarth B, Bouchard C: The human gene map for performance and health-related fitness phenotypes: the 2003 update. Med Sci Sports Exerc 2004;36: 1451–1469.

4 Wolfarth B, Bray MS, Hagberg JM, Pérusse L, Rauramaa R, Rivera MA, Roth SM, Rankinen T, Bouchard C: The human gene map for performance and health-related fitness phenotypes: the 2004 update. Med Sci Sports Exerc 2005;37:881–903.

5 Rankinen T, Bray MS, Hagberg JM, Pérusse L, Roth SM, Wolfarth B, Bouchard C: The human gene map for performance and health-related fitness phenotypes: the 2005 update. Med Sci Sports Exerc 2006; 38:1863–1888.

6 Bray MS, Hagberg JM, Pérusse L, Rankinen T, Roth SM, Wolfarth B, Bouchard C: The human gene map for performance and health-related phenotypes: the 2006–2007 update. Med Sci Sports Exerc 2009;41: 35–73.

7 Rankinen T, Roth SM, Bray MS, Loos R, Pérusse L, Wolfarth B, Hagberg JM, Bouchard C: Advances in exercise, fitness, and performance genomics. Med Sci Sports Exerc 2010;42:835–846.

8 Hagberg JM, Rankinen T, Loos RJ, Pérusse L, Roth SM, Wolfarth B, Bouchard C: Advances in exercise, fitness, and performance genomics in 2010. Med Sci Sports Exerc 2011;43:743–752.

9 Roth SM, Rankinen T, Hagberg JM, Loos RJ, Pérusse L, Sarzynski MA, Wolfarth B, Bouchard C. Advances in exercise, fitness, and performance genomics in 2011. Med Sci Sports Exerc 2012;44:809–817.

10 Pérusse L, Rankinen T, Hagberg JM, Loos RJ, Roth SM, Sarzynski MA, Wolfarth B, Bouchard C: Advances in exercise, fitness, and performance genomics in 2012. Med Sci Sports Exerc 2013;45:824–831.

11 Wolfarth B, Rankinen T, Hagberg JM, Loo, RJ, Pérusse L, Roth SM, Sarzynski MA, Bouchard C: Advances in exercise, fitness, and performance genomics in 2013. Med Sci Sports Exerc 2014;46:851–859.

12 Loos RJ, Hagberg JM, Pérusse L, Roth SM, Sarzynski MA, Wolfarth B, Rankinen T, Bouchard C: Advances in exercise, fitness, and performance genomics in 2014. Med Sci Sports Exerc 2015;47:1105–1112.

13 De Moor MH, Spector TD, Cherkas LF, Falchi M, Hottenga JJ, Boomsma DI, De Geus EJ: Genome-wide linkage scan for athlete status in 700 British female DZ twin pairs. Twin Res Hum Genet 2007;10: 812–820.

14 Bouchard C: Genomic predictors of trainability. Exp Physiol 2012;97:347–352.

15 Stubbe JH, Boomsma DI, Vink JM, Cornes BK, Martin NG, Skytthe A, Kyvik KO, Rose RJ, Kujala UM, Kaprio J, Harris JR, Pedersen NL, Hunkin J, Spector TD, de Geus EJ: Genetic influences on exercise participation in 37,051 twin pairs from seven countries. PLoS One 2006;1:e22.

16 Rivera MA, Pérusse L, Simoneau JA, Gagnon J, Dionne FT, Leon AS, Skinner JS, Wilmore JH, Province M, Rao DC, Bouchard C: Linkage between a muscle-specific CK gene marker and VO_{2max} in the HERITAGE Family Study. Med Sci Sports Exerc 1999;31:698–701.

17 Bouchard C, Rankinen T, Chagnon YC, Rice T, Pérusse L, Gagnon J, Borecki I, An P, Leon AS, Skinner JS, Wilmore JH, Province M, Rao DC: Genomic scan for maximal oxygen uptake and its response to training in the HERITAGE Family Study. J Appl Physiol 2000;88:551–559.

18 Rankinen T, Pérusse L, Borecki I, Chagnon YC, Gagnon J, Leon AS, Skinner JS, Wilmore JH, Rao DC, Bouchard C: The Na(+)-K(+)-ATPase alpha2 gene and trainability of cardiorespiratory endurance: the HERITAGE family study. J Appl Physiol 2000;88: 346–351.

19 Rico-Sanz J, Rankinen T, Rice T, Leon AS, Skinner JS, Wilmore JH, Rao DC, Bouchard C: Quantitative trait loci for maximal exercise capacity phenotypes and their responses to training in the HERITAGE Family Study. Physiol Genomics 2004;16:256–260.

20 Rankinen T, An P, Pérusse L, Rice T, Chagnon YC, Gagnon J, Leon AS, Skinner JS, Wilmore JH, Rao DC, Bouchard C: Genome-wide linkage scan for exercise stroke volume and cardiac output in the HERITAGE Family Study. Physiol Genomics 2002;10:57–62.

21 Simonen RL, Rankinen T, Pérusse L, Rice T, Rao DC, Chagnon Y, Bouchard C; Quebec Family Study: Genome-wide linkage scan for physical activity levels in the Quebec Family Study. Med Sci Sports Exerc 2003;35:1355–1359.

22 Cai G, Cole SA, Butte N, Bacino C, Diego V, Tan K, Göring HH, O'Rahilly S, Farooqi IS, Comuzzie AG: A quantitative trait locus on chromosome 18q for physical activity and dietary intake in Hispanic children. Obesity (Silver Spring) 2006;14:1596–1604.

23 Huygens W, Thomis MA, Peeters MW, Aerssens J, Janssen R, Vlietinck RF, Beunen G: Linkage of myostatin pathway genes with knee strength in humans. Physiol Genomics 2004;17:264–270.

24 Huygens W, Thomis MA, Peeters MW, Aerssens J, Vlietinck R, Beunen GP: Quantitative trait loci for human muscle strength: linkage analysis of myostatin pathway genes. Physiol Genomics 2005;22: 390–397.
25 MacArthur DG, North KN: Genes and human elite athletic performance. Hum Genet 2005;116:331–339.
26 Pitsiladis Y, Wang G, Wolfarth B, Scott R, Fuku N, Mikami E, He Z, Fiuza-Luces C, Eynon N, Lucia A: Genomics of elite sporting performance: what little we know and necessary advances. Br J Sports Med 2013;47:550–555.
27 Wang G, Padmanabhan S, Wolfarth B, Fuku N, Lucia A, Ahmetov II, Cieszczyk P, Collins M, Eynon N, Klissouras V, Williams A, Pitsiladis Y: Genomics of elite sporting performance: what little we know and necessary advances. Adv Genet 2013;84:123–149.
28 Guth LM, Roth SM: Genetic influence on athletic performance. Curr Opin Pediatr 2013;25:653–658.
29 Wagner JK, Royal CD: Field of genes: an investigation of sports-related genetic testing. J Pers Med 2012;2:119–137.
30 Yang N, MacArthur DG, Gulbin JP, Hahn AG, Beggs AH, Easteal S, North K: *ACTN3* genotype is associated with human elite athletic performance. Am J Hum Genet 2003;73:627–631.
31 Webborn N, Williams A, McNamee M, Bouchard C, Pitsiladis Y, Ahmetov I, Ashley E, Byrne N, Camporesi S, Collins M, Dijkstra P, Eynon N, Fuku N, Garton FC, Hoppe N, Holm S, Kaye J, Klissouras V, Lucia A, Maase K, Moran C, North KN, Pigozzi F, Wang G: Direct-to-consumer genetic testing for predicting sports performance and talent identification. Br J Sports Med 2015;49:1486–1491.
32 Bouchard C: Exercise genomics – a paradigm shift is needed: a commentary. Br J Sports Med 2015;49: 1492–1496.
33 Pitsiladis YP, Tanaka M, Eynon N, Bouchard C, North KN, Williams AG, Collins M, Moran CN, Britton SL, Fuku N, Ashley EA, Klissouras V, Lucia A, Ahmetov II, de Geus E, Alsayrafi M; Athlome Project Consortium: The Athlome Project Consortium: a concerted effort to discover genomic and other 'omic' markers of athletic performance. Phys Genomics 2016;48:183–190.
34 Rankinen T, Fuku N, Wolfarth B, Wang G, Sarzynski MA, Alexeev DG, Ahmetov II, Boulay MR, Cieszczyk P, Eynon N, Filipenko ML, Garton FC, Generozov EV, Govorun VM, Houweling PJ, Kawahara T, Kostryukova ES, Kulemin NA, Larin AK, Maciejewska-Karłowska A, Miyachi M, Muniesa CA, Murakami H, Ospanova EA, Padmanabhan S, Pavlenko AV, Pyankova ON, Santiago C, Sawczuk M, Scott RA, Uyba VV, Yvert T, Perusse L, Ghosh S, Rauramaa R, North KN, Lucia A, Pitsiladis Y, Bouchard C: No evidence of a common DNA variant profile specific to world class endurance athletes. PLoS One 2016;11:e0147330.

Prof. Yannis P. Pitsiladis
FIMS Reference Collaborating Centre of Sports Medicine for Anti-Doping Research
University of Brighton
Welkin House, 30 Carlisle Road
Eastbourne, BN20 7SN (UK)
E-Mail y.pitsiladis@brighton.ac.uk

Posthumus M, Collins M (eds): Genetics and Sports, ed 2, revised, extended.
Med Sport Sci. Basel, Karger, 2016, vol 61, pp 68–91 (DOI: 10.1159/000445243)

Genes and Musculoskeletal Soft-Tissue Injuries

Masouda Rahim · Malcolm Collins · Alison September

Division of Exercise Science and Sports Medicine, Department of Human Biology, Faculty of Health Sciences, University of Cape Town, Cape Town, South Africa

Abstract

There is mounting evidence suggesting a genetic contribution to the susceptibility of sustaining a musculoskeletal soft-tissue injury. To date, more than 70 loci have been implicated in several injury profiles. The genes implicated through these loci encode a broad spectrum of matrix proteins including collagens and non-collagens. The large majority of these studies have followed a candidate gene case-control study design. A small proportion of these loci have been repeated in independent studies, of which some have included different musculoskeletal injuries. However, the large majority of these studies are underpowered to detect contributions of small effect sizes (odds ratio <2.0). It is therefore critical that large data sets are collected and that consortia are established to effectively pool resources to understand the biological significance of these genetic loci and risk susceptibility. We are in the era of *omics* and *high-throughput* technologies but it is only through collaborations that we will realize the clinical significance of the genomic revolution and its application to musculoskeletal soft-tissue injury susceptibility.

Musculoskeletal soft tissues such as tendon, ligament and skeletal muscle are commonly injured as a result of participation in regular and/or repetitive physical activity within either a recreational or occupational environment [1]. The aetiology of these injuries remains largely uncharacterized with several risk factors being suggested although much debate exists on the level of evidence for the proposed risk factors [2]. Familial and case-control association studies have provided evidence

Table 1. Genes encoding extracellular matrix proteins implicated with susceptibility to musculoskeletal soft-tissue injuries

Candidate gene	Encoded protein	Protein function	Chromosomal location	Polymorphism	Location	MSK injury	Ref.
Structural components							
COL1A1	α_1(I) collagen chain	Major fibrillar collagen	17q21.33	**rs1107946 (–1997 G/T)**	Promoter	ACL[a]	8
				rs1800012 (+1245 G/T; Sp1)	Intron 1	ACL[b], shoulder dislocations, acute soft-tissue ruptures, LDD	4–9, 11, 12
COL3A1	α_1(III) collagen chain	Major fibrillar collagen	2q31	**rs1800255 (G/A; A198T)**	Exon 30	ACL	16, 17
				rs2056156 (C/T)	Intron 1	LDD	18
				rs2203601 (A/T)	Intron 10		
COL5A1	α_1(V) collagen chain	Minor fibrillar collagen	9q34.2–q34.3	rs13945 (C/T; *Dpn*II RFLP)	3′-UTR	CTS	22
				rs12722 (T/C; *BstU*I RFLP)		AT, ACL, CTS	16, 19–22
				rs3196378 (C/A *Aci*I RFLP)		AT[b]	20
				rs71746744 (–/AGGG)		AT	23
				rs16399 (ATCT/–)			
				rs1134170 (A/T)			
COL9A2	α_2(IX) collagen chain	Major collagen component of hyaline cartilage	1p33–p32	rs209923 (–2066 C/A)	Promoter	LDD[a]	30
				rs7533552 (+977 A/G)	Exon 19		
				rs2076696 (+1147 G/C)	Intron 19		
				rs12077871 (Trp2; +976 C/T; R326W)	Exon 19	LDD	26–28, 31
COL9A3	α_3(IX) collagen chain		20q13.3	rs61734651 (C/T; Trp3; R103W)	Exon 5	LDD	32, 33
COL11A1	α_1(XI) collagen chain	Minor fibrillar collagen	1p21	**rs3753841 (T/C; L1323P)**	Exon 52	AT[a]	37
				rs1676486 (C/T; P1535S)	Exon 62	AT[a], LDH	36, 37
COL11A2	α_2(XI) collagen chain		6p21.3	**rs1799907 (IVS6 –4 T/A)**	Intron 6	AT[a]	37
COL12A1	α_1(XII) collagen chain	FACIT collagen	6q12–q13	rs970547 (A/G; *Alu*I RFLP; S3058G)	Exon 65	ACL	16, 38
COL27A1	α_1(XXVII) collagen chain	Minor fibrillar collagen	9q32	**rs946053 (G/T)**	Intron 41	AT[a]	40
TNC	Tenascin-C glycoprotein	Regulates cell-matrix interactions	9q33	G-T tandem repeat	Intron 17	AT	39
				rs1330363 (C/T)	Intron 15	AT[a]	40
				rs2104772 (T/A; L1677I)	Exon 17		
				rs13321 (G/C; E2008Q)	Exon 24		
THBS2	Thrombospondin-2	Mediates cell-cell and cell-matrix interactions	6q27	**rs9406328 (IVS10 –8 C/T)**	Intron 10	LDH	41
FBN2	Fibrillin-2	Component of connective tissue microfibrils	5q23–q31	rs331079 (G/T)	Intron 7	AT, ACL	42

Table 1. Continued

Candidate gene	Encoded protein	Protein function	Chromosomal location	Polymorphism	Location	MSK injury	Ref.
CILP	Cartilage intermediate layer protein	Interacts directly with transforming growth factor-β_1	15q22	**rs2073711 (+1184 T/C; I395T)**	Exon 8	LDD	43, 44
Regulators of the ECM							
MIR608	MicroRNA 608	Regulation of *COL5A1* mRNA stability	10q24.31	rs4919510 (C/G)	3'-UTR	AT	23
Proteoglycans							
ACAN	Aggrecan	Major proteoglycan of articular cartilage – contributes to load-bearing properties of cartilage	15q26.1	rs2351491 (C/T) rs1042631 (C/T) rs1516797 (T/G) **VNTR**	Exon 11 Intron 12 Exon 12	ACL[a] ACL[a], LDD ACL[b], LDD LDD, LDH	57 18, 57 52–56
BGN	Biglycan	Involved in collagen fibrillogenesis	Xq28	rs1126499 (+189 C/T) rs1042103 (+359 G/A)	Exon 4 Exon 8	ACL[a], CTS ACL[a]	57, 58 57
DCN	Decorin		12q21.33	rs13312816 (C/T) rs516115 (T/G)	Intron 1 Intron 3	ACL[a] ACL[b]	57
LUM	Lumican		12q21.33	rs2268578 (+697 T/C)	Intron 3	ACL[a]	57
ASPN	Asporin	Cartilage ECM protein	9q22	**rs387906276 [14 Asp (D) repeat]**	Intron	LDD	61
Matrix proteinases							
MMP1	Matrix metalloproteinase-1	Degradation of collagenous and non-collagenous ECM components	11q22.3	**rs1799750 (–1607 1G/2G)** **rs1144393 (–519 A/G)**	Promoter	ACL[a], LDD, PTT[b] PTT[b]	67–69 72
MMP2	Matrix metalloproteinase-2		16q13–q21	**rs243865 (–1306 C/T)** **rs2285053 (–735 C/T)**	Promoter	LDD	77, 79
MMP3	Matrix metalloproteinase-3		11q22.3	**rs3025058 (–1612 5A/6A)** **rs679620 (A/G; E45K)** rs591058 (+1547 T/C) rs650108 (+495 A/G)	Promoter Exon 2 Intron 4 Intron 8	ACL, LDD ACL[a], ATP[b] ATP[b]	84–86 67, 82 82
MMP8	Matrix metalloproteinase-8		11q22.3	**rs11225395 (–799 C/T)**	Promoter	PTT	73
MMP9	Matrix metalloproteinase-9		20q11.2–q13.1	**rs3918242 (–1562 C/T)** **rs17576 (G/A; Q279R)**	Promoter Exon 6	LDD LDH	80 41
MMP10	Matrix metalloproteinase-10		11q22.3	rs486055 C/T (R53K)	Exon 1	ACL[a]	67
MMP12	Matrix metalloproteinase-12		11q22.3	**rs2276109 (–82 A/G)**	Promoter	ACL[b]	67

Table 1. Continued

Candidate gene	Encoded protein	Protein function	Chromosomal location	Polymorphism	Location	MSK injury	Ref.
ADAMTS5	ADAM with thrombospondin type 1 motif, 5	Cleaves aggrecan	21q21.3	rs1652502 (C/T) rs229052 (A/G) rs151058 (A/G)	Intron 1 Intron 2 Intron 3	LDD[a]	87
TIMP2	Metalloproteinase inhibitor-2	Essential regulator of ECM turnover and remodelling	17q25	rs4789932 (–2803 C/T)	Promoter	ATP	89
Cytokines and signalling factors							
IL1A	IL-1α	Role in the inflammatory pathway	2q14	**rs1800587 (–889 G/A)**	5′-UTR	LDD	95
IL1B	IL-1β	Role in the inflammatory pathway and ECM degradation	2q14	**rs16944 (–511 T/C)** **rs1143627 (–31 C/T)**	Promoter	AT[a]	96
IL1RN	IL-1 receptor antagonist	Antagonist for IL-1α and IL-1β	2q14.2	**rs2234663 VNTR**	Intron 2	AT[a]	96
IL6	IL-6	Role in apoptosis and the inflammatory pathway	1q21	**rs1800795 (–172 G/C)**	Promoter	AT[a], LDD[b]	44, 96
IL6R	IL-6 receptor	Receptor for IL-6	1q21	**rs2228145 (A/C; D358A)**	Exon 9	CTS[b]	107
IL10	IL-10	Anti-inflammatory cytokine	1q31–q32	**rs1800896 (–1082 A/G)** **rs1800872 (–592 A/C)**	Promoter	LDD	109
IL18R1	IL-18 receptor-1	Receptor for IL-18	2q12.1	**rs2287037 (–94 A/G)**	Promoter	LDD	112
IL18RAP	IL-18 receptor accessory protein	Enhances IL-18 binding to IL-18R1 and involved in IL-18 signalling	2q12.1	rs1420106 (T/C) rs1420100 (T/G) **rs917997 (A/G)**	Promoter Intron 3′-UTR	LDD	18, 112, 114
VEGFA	VEGF-A isoform	Essential regulator of angiogenesis	6p21.1	**rs699947 (–2578 C/A)** **rs1570360 (–1154 G/A)** **rs2010963 (–634 G/C)**	Promoter Promoter 5′-UTR	ACL[b]	120
KDR	Kinase insert domain receptor	Receptor for VEGF and mediates VEGF signalling	4q11–4q12	**rs2071559 (–604 G/A)** **rs1870377 (+1719 T/A; H472Q)**	Promoter Exon 11	ACL[b]	120
HIF1A	Hypoxia-inducible factor-1, α-subunit	Regulates cellular response to hypoxia	14q23.2	**rs11549467 (+1790 G/A; A588T)**	Exon 12	LDD	126
GDF5	Growth differentiation factor-5	Regulates cell growth and differentiation	20q11	**rs143383 (T/C)**	5′-UTR	ATP, LDD	132, 133
VDR	Vitamin D receptor	Regulates several metabolic pathways (incl. immune response)	12q13.11	**rs2228570 (C/T; FokI RFLP; M1T)** rs731236 (+1056 T/C; *TaqI* RFLP; T352C)	5′-UTR 3′-UTR	LDD LDD	136 54, 136–139

Table 1. Continued

Candidate gene	Encoded protein	Protein function	Chromosomal location	Polymorphism	Location	MSK injury	Ref.
Apoptosis							
CASP8	Caspase-8	Initiator caspase	2q33–q34	**rs3834129 ins/del (–652 6N del)** **rs1045485 G/C (D302H)**	Promoter Exon 9	AT[b]	144
CASP9	Caspase-9	Initiator caspase	1p36.21	**rs1052576 (Ex5+32 G/A; Q221R)**	Exon 5	LDD	149
FASL	Fas ligand	Induces apoptosis upon binding to Fas receptor	1q23	**rs763110 (–844 C/T)**	Promoter	LDD	153
FAS	Fas cell surface death receptor	Receptor for Fas ligand	10q24.1	**rs2234767 (–1377 G/A)**	Promoter	LDD	153
BCL2	B-cell lymphoma-2	Regulates apoptosis, anti-apoptotic	18q21.3	**rs2279115 (–938 C/A)**	Promoter	LDD	156

The polymorphisms in bold indicate functional or potentially functional variants. ACL = Anterior cruciate ligament ruptures; AT = Achilles tendinopathy; ATP = Achilles tendon pathology; CTS = carpal tunnel syndrome; ECM = extracellular matrix; FACIT = fibril-associated collagens with interrupted triple helices; IL = interleukin; LDD = lumbar disc disease; LDH = lumbar disc herniation; MSK = musculoskeletal; PTT = tendinopathy of the posterior tibialis tendon; RFLP = restriction fragment length polymorphism; VEGF = vascular endothelial growth factor; VNTR = variable nucleotide tandem repeat.
[a] The variant was associated as part of an inferred haplotype or an inferred allele combination.
[b] The variant was independently associated with the MSK injury, as well as part of an inferred haplotype or an inferred allele combination.

suggesting that a genetic susceptibility may contribute to the interindividual variation in response to mechanical load, which may consequently affect tissue integrity, function and ultimately injury susceptibility. Several genetic regions (loci) within candidate genes have been investigated, and several sequence variants have been implicated (table 1). This review will focus on summarizing these genetic loci and describe some of the proposed biological mechanisms underpinning these associations.

The genes implicated (table 1) encode both collagenous and non-collagenous matrix constituents, and they broadly function primarily as (i) structural components of the musculoskeletal system which include the collagens and glycoproteins and (ii) regulators of the extracellular matrix (ECM) such as the proteoglycans, matrix metalloproteinases, tissue inhibitors of matrix metalloproteinases, cytokines, growth factors and caspases to name a few. There is evidence to suggest that these matrix components may individually or collectively have key functional roles in processes such as fibrillogenesis, cell signalling, apoptosis, angiogenesis, hypoxia, healing, wound repair and pain signalling. An overlap between some of the genes implicated in these genetic association studies and protein or mRNA expression profiling studies have been noted.

Structural Components

Several collagen-encoding genes have been implicated with the risk of several connective tissue disorders and/or musculoskeletal injuries. Type I collagen, a major fibrillar collagen, is the most abundant protein within connective tissue, and the risk of several musculoskeletal injuries has been linked to sequence variants mapped to the gene encoding the α_1(I)-chain *(COL1A1).* The functional Sp1 binding site polymorphism (rs1800012) within intron 1 encodes a G-to-T substitution resulting in increased binding affinity for Sp1 and consequently greater *COL1A1* transcription [3]. This polymorphism has been associated with cruciate ligament ruptures [4–6], shoulder dislocations [5] and acute soft-tissue ruptures in a combined analysis [7]. An under-representation of the TT genotype in the cases compared to the controls was consistently reported, indicating a possible reduced risk of injury [5–9]. However, the underlying biological mechanisms require further investigation. Conversely, the TT genotype was associated with an increased risk of anterior cruciate ligament (ACL) ruptures in a study investigating Polish skiers [4]; however, it should be noted that skiing involves a different mechanism of injury [10]. The TT genotype was also associated with an increased risk of disc degeneration [11] and lumbar disc disease (LDD) [12]. Recently, a second functional *COL1A1* promoter polymorphism (–1997 G/T; rs1107946) was investigated with the risk of ACL ruptures in professional soccer players [8]. The –1997 G/T variant is in linkage disequilibrium with the Sp1 polymorphism (+1245 G/T), and together they form a functional haplotype that regulates transcription [13]. Although not independently associated in this study, probably due to sample size, the –1997 G/T and +1245 G/T (Sp1) G-T inferred haplotype was linked with a reduced risk of ACL ruptures [8].

Type III collagen is another major fibrillar collagen, and increased type III collagen is a characteristic of tendinopathic tendons [14]. The non-synonymous rs1800255 G/A (Ala531Thr) variant, within the *COL3A1* gene encoding the α_1(III)-chain, is suggested to affect collagen fibre assembly and tensile strength [15]. In 2 independent studies, the rs1800255 AA genotype was associated with an increased risk of ACL ruptures in a study group from Poland [16, 17]. In addition, a *COL3A1* rs2056156 and rs2203601 inferred haplotype was associated with lumbar disc bulging [18].

Several sequence variants (rs13946 T/C, rs14776422/rs5574880 W/M, rs12722 T/C, rs3196378 C/A, rs71746744 –/AGGG, rs16399 –/ATCT, rs1134170 A/T) within the 3′-untranslated region (3′-UTR) of the *COL5A1* gene have also been independently, or in combination, associated with several musculoskeletal injuries (table 1). *COL5A1* codes for the α_1(V)-chain of type V collagen, a minor fibrillar collagen involved in the regulation of collagen fibril diameter and lateral growth during fibrillogenesis. To date, this genomic interval has been associated with Achilles tendinopathy (AT) [19, 20], ACL ruptures in females [16, 21] and carpal tunnel syndrome (CTS) [22]. Additionally, the *MIR608* rs4919510 CC genotype was independently associated with an increased risk of AT [23]. *MIR608* encodes a 25-bp microRNA which binds to the Hsa-miR-608 binding site in the *COL5A1* 3′-UTR.

Several of the studied *COL5A1* polymorphisms appear to alter the predicted secondary structure of the 3′-UTR which may impact *COL5A1* mRNA stability [23]. The combination of 7 tightly linked polymorphisms located within the 3′-UTR constitute 2 major functional forms, termed the 'C functional form' and the 'T functional form' which have been linked with a decreased and increased risk of AT, respectively [24]. Moreover, the T functional form was associated with an overall increase in mRNA stability [24]. Collins and Posthumus [25] hypothesized that altered *COL5A1* mRNA stability results in altered type V collagen production, and changes in collagen fibril diameter and density, thus potentially altering the biomechanical properties of musculoskeletal tissues.

Type IX collagen is a major collagen component of hyaline cartilage and is thought to serve as a bridge between collagenous and non-collagenous proteins in tissues [26]. Structurally, it is a heterotrimeric protein consisting of 3 chains: α_1(IX), α_2(IX) and α_3(IX), encoded by the *COL9A1*, *COL9A2* and *COL9A3* genes, respectively. Genetic variants within the *COL9A2* and *COL9A3* genes have been associated with LDD risk (table 1), although ethnic differences in allele frequency have been noted [26–30]. Trp2, a *COL9A2* allele that replaces the wild-type arginine with tryptophan at codon 326 (rs12077871 C/T), was associated with intervertebral disc disease and corresponded with increased degeneration [26–28, 31]. Additionally, a *COL9A2* inferred haplotype (A-G-C) consisting of the rs209923 (–2066 C/A), rs7533552 (+977 A/G) and rs2076696 (+1147G/C) variants was overrepresented in a Japanese LDD group [30]. The *COL9A3* Trp3 allele, encoding an arginine-to-tryptophan (R103W) substitution, was also associated with LDD [32, 33], and the presence of at least 1 Trp3 allele increased the risk of LDD approximately 3-fold [32].

Type XI collagen is a minor fibrillar collagen that is structurally and functionally related to type V collagen [34]. It is predominately expressed in cartilage as well as in developing tendons [35]. Several single-nucleotide polymorphisms (SNPs) within the genes encoding the α_1(XI)- and α_2(XI)-chains (*COL11A1* and *COL11A2*, respectively) have been associated with the risk of LDD and AT. *COL11A1* mRNA was substantially expressed in intervertebral discs, and the expression level was inversely correlated with the severity of disc degeneration [36]. The rs1676486 (4603 C/T; Pro1535Ser) T allele was associated with lumbar disc herniation (LDH) in 3 independent studies from Japan [36]. Moreover, the 4603 T allele had significantly lower expression levels than the C allele and also appears to be associated with reduced mRNA stability [36]. Hay et al. [37] also explored variants within the type XI collagen genes with a risk of AT: specifically *COL11A1* rs3753841 (3968 T/C; Leu1323Pro), rs1676486 C/T and *COL11A2* rs1799907 (IVS6 –4 T/A). The rs3753841 residue change is predicted to be deleterious, and the rs1799907 T allele results in alternative splicing of the *COL11A2* gene. Although not independently associated, the *COL11A1*-*COL11A2* inferred allele combination was associated with AT. The T-C-T inferred haplotype (rs3753841 T/C; rs1676486 C/T; rs17999079 T/A) conferred an increased risk of AT [37]. Moreover, the addition of the previously associated *COL5A1* rs71746744 (–/AGGG) polymorphism [23] to the inferred haplotype showed an overrepresentation of the T-C-T-(AGGG) combination in the AT group. The authors hypothesized that the genes may potentially interact to regulate collagen fibril assembly and modulate the risk of AT [37]. Interestingly, the associated haplotype consisted of the alternate alleles to those associated with the risk of LDH. A similar observation was noted at the *COL1A1* Sp1 binding site locus with the TT genotype conferring an increased risk of LDD but associated with reduced risk of ruptures of the ACL, AT and shoulder dislocations [5–8].

Collagen type XII is a homotrimer consisting of 3 α_1(XII)-chains, coded for by the *COL12A1* gene. The protein is a member of the fibril-associated collagen with interrupted triple helix subfamily and functionally has a similar role to type V collagen during fibrillogenesis. The rs970547 A/G SNP in exon 65, also referred to as the *Alu*I restriction fragment length polymorphism (RFLP), encodes an amino acid change from serine to glycine at position 3058 (S3058G) with unknown functional effects. The rs970547 SNP was found to modulate the risk of ACL ruptures in females, with the AA genotype conferring a 2.4-fold greater risk [38]. A repeat study by O'Connell et al. [16] investigated the *Alu*I RFLP in a larger South African group and in a Polish group. The AA genotype was once again significantly overrepresented in the female ACL participants compared to the female controls in the Polish group. In addition, the authors explored the interaction between the *COL5A1* rs12722 T/C and the *COL12A1* rs970547 A/G variants and found the T-A inferred allele combination to be significantly overrepresented in the female ACL group compared to the female controls in both populations.

The glycoprotein, tenascin-C, forms a key component of tendons and ligaments and also plays a role in regulating cell-matrix interactions. The *TNC* gene, mapped to chromosome 9q33, contains a guanine-thymine (GT) dinucleotide repeat in intron 17. Alleles with 12 and 14 GT repeats were overrepresented in the injured group whereas alleles with 13 and 17 GT repeats were underrepresented in the injured group [39]. A study by Saunders et al. [40] investigated additional polymorphisms within the *TNC* gene with the risk of AT in 2 populations. Although no independent genotype associations were noted, allele and inferred haplotype effects were observed. The rs2104772 T allele was associated with a decreased risk of AT and the rs1330363 G allele with an increased risk. Moreover, the authors investigated variants within the neighbouring *COL27A1* gene, which lies 708 kb upstream of the *TNC* gene, in an effort to narrow the genetic interval harbouring potential risk variants for AT. *COL27A1* codes for the α_1(XXVII)-chain of type XXVII collagen, an atypical fibrillar collagen. The authors explored potential haplotypes using the most distal *COL27A1* variant (rs946053 G/T) and the 2 proximal *TNC* variants (rs13321 G/C and rs2104772 T/A), and found that the G-C-A inferred haplotype was associated with an increased risk of AT in both a South African and Australian group.

Thrombospondin-2 *(THBS2)* is an ECM glycoprotein that mediates cell-cell and cell-matrix interactions and also regulates effective levels of matrix metalloproteinase (MMP)-2 and -9. The intronic rs9406328 (IVS10 –8 C/T) SNP was associated with LDH in 2 independent study groups from Japan, with the TT genotype overrepresented in the cases [41]. Further analysis revealed the polymorphism results in alternative splicing of the *THBS2* gene: the TT genotype displayed a higher rate of exon 11 skipping which consequently reduces the binding affinity of thrombospondin-2 to MMP-2 and MMP-9 [41].

Fibrillin-2, encoded by the *FBN2* gene, is a component of connective tissue microfibrils. The intronic rs331079 G/T SNP has recently been associated with AT and ACL ruptures; specifically, the GG genotype was significantly overrepresented within the AT group compared to the controls, and the G allele was also significantly different between the AT group and the ACL group compared to the controls [42].

Cartilage intermediate layer protein interacts directly with transforming growth factor-β_1 (TGF-β_1) and can also inhibit the TGF-β_1-mediated induction of ECM proteins. Cartilage intermediate layer protein is expressed abundantly in intervertebral discs and increasingly expressed as degeneration progresses [43]. A functional polymorphism in exon 8 (rs2073711; 1184 T/C; I395T) was associated with an increased risk of LDD (1184 C allele) and also showed increased inhibition of TGF-β_1 in a study group from Japan [43]. However, in a study group from Finland it was the 1184 T allele that was found to be associated with the LDD risk [44].

Regulators of the Extracellular Matrix

Proteoglycans

Proteoglycans are important structural constituents of the ECM with key roles in resistance of compressive forces, collagen fibrillogenesis, matrix remodelling and cell signalling [45, 46]. Corps et al. [47, 48] identified altered proteoglycan mRNA levels in chronic painful and ruptured Achilles tendons compared to healthy tendons. Specifically, aggrecan and biglycan expression was increased in degenerative tendons whereas versican levels were decreased. In ruptured tendons, both versican and decorin mRNA expression levels were significantly decreased compared to healthy tendons. Young et al. [49] reported decreased biglycan, decorin and versican protein levels in ruptured compared to non-ruptured ACLs. Interestingly, Johnson et al. [50] identified significantly higher gene expression levels of aggrecan and fibromodulin in female ruptured ACL tissue compared to male tissue.

Genetic variants within the genes encoding proteoglycans have been investigated with the risk of musculoskeletal injuries. A variable nucleotide tandem repeat, located in exon 12 of the aggrecan-encoding gene (*ACAN)*, codes for the chondroitin sulphate attachment domain. This polymorphism results in varying lengths of the aggrecan core protein and is considered to have an effect on the functional properties of the cartilage [51], with the shorter variable nucleotide tandem repeat length found to be associated with multilevel and severe disc degeneration [52–56]. A study by Mannion et al. [57] investigated several genetic variants within genes encoding proteoglycans with the risk of ACL ruptures. Three polymorphisms within *ACAN* were investigated both independently and as a haplotype. The rs1516797 G allele was significantly underrepresented in the control group compared to the ACL group, and the genomic interval containing the *ACAN* rs2351491 C/T, rs1042631 T/C and rs1516797 T/G polymorphisms was implicated with ACL injuries. Moreover, the rs1042631 and rs1516797 SNPs were implicated with LDD [18] but no association was noted with the rs1516797 variant and the risk of CTS [58]. The biological significance of these polymorphisms is unknown.

Biglycan, encoded by the *BGN* gene localized on chromosome Xq28, functions as a key structural and signalling molecule. As a structural component of the ECM, biglycan is involved in collagen fibrillogenesis and regulation of bone formation [46]. This small leucine-rich proteoglycan also plays a role in inflammation and acts as a modulator of several growth factor and cytokine functions, including TGF-β_1 [59]. Although not independently associated, the *BGN* rs1126499 C/T and rs1042103 G/A inferred C-G haplotype was implicated with a decreased risk of ACL injury in female participants [57]. Furthermore, the rs1126499 CC genotype was associated with a decreased risk of CTS [58]. Decorin is the most abundant proteoglycan in tendon tissue, and a marked decrease in expression levels of decorin mRNA was observed in ruptured tendons whereas there were no significant differences between painful tendons compared to healthy tendons [47]. Polymorphisms within *DCN* were also implicated in ACL risk model analyses. More specifically, the *DCN* rs516115 GG genotype was overrepresented 9-fold in the female control group compared to the ACL group. In addition, the AA genotype was significantly associated with an increased risk of non-contact ACL ruptures in female participants [57]. Haplotype analyses further implicated genomic intervals spanning the lumican- and decorin-encoding genes (*LUM* rs2268578 T/C, *DCN* rs13312816 A/T and *DCN* rs516115 A/G) with ACL injury susceptibility [57].

Asporin, though not a proteoglycan, is an ECM protein belonging to the leucine-rich repeat family and is closely related to biglycan and decorin. The protein is encoded by the *ASPN* gene and contains a unique stretch of aspartate (D) residues in the N terminus, a polymorphic region of the gene [60]. *ASPN* expression was upregulated in degenerative discs, and individuals with the D14 allele (14 aspartate residues) were found to be at an increased risk of LDD in 2 Asian populations [61].

Matrix Proteinases

MMPs are a large family of zinc-dependent endopeptidases involved in the degradation of collagenous and non-collagenous matrix components [62]. They play key roles in tissue degradation after mechanical loading to allow generation of new tissue and are therefore critical to matrix integrity and homeostasis [63]. Recent insights suggest a role for MMPs in cell proliferation, differentiation, angiogenesis and the immune system [62, 63]. Increased MMP expression has been reported in tendinopathic tissues as well as in ruptured tendon and ligament tissues [64–66].

The 11q22 chromosomal locus harbours a cluster of *MMP* genes, several of which have been associated with musculoskeletal injuries. An inferred allele combination of genetic variants in the *MMP10, MMP1, MMP3* and *MMP12* genes (rs486055 C/T, rs1799750 1G/2G, rs679620 A/G and rs2276109 A/G, respectively) was associated with a decreased risk of ACL injuries [67]. Additionally, an independent association was noted with the functional *MMP12* rs2276109 AA genotype conferring an increased risk of non-contact ACL ruptures [67]. The functional *MMP1* promoter polymorphism –1607 1G/2G (rs1799750) was implicated in LDD [68] and with tendinopathy of the posterior tibialis tendon (PTT) [69]. The 2G allele, which is linked with increased *MMP1* transcription compared to the 1G allele [70, 71], was reported to be overrepresented in the PTT tendinopathy group [69]. A second functional promoter polymorphism at position –519 (rs1144393 A/G) was also independently associated with PTT tendinopathy, with the G allele conferring a greater risk, and as a haplotype with the –1607 1G/2G polymorphism [72]. In an earlier study by the same authors, the functional *MMP8* (collagenase-2) –799 C/T polymorphism (rs11225395) was implicated in PTT dysfunction: both the TT genotype and T allele were significantly associated with an increased risk of developing tendinopathy [73]. The –799 T allele was previously shown to be linked with decreased transcription factor binding [74].

MMP-2 and MMP-9 have been shown to have increased activity and mRNA expression in both degenerative Achilles tendons and degenerative discs [64, 75, 76]. Two polymorphisms in the promoter region of *MMP2* have been linked with LDD: the –1306 T allele disrupts an Sp1 binding site resulting in lower transcriptional activity whereas the C allele of the same SNP is linked with increased transcription [77], and the –735 T allele is linked with lower promoter activity [78]. Individuals with the –1306 CC genotype had a 3-fold increased risk for LDD [77], and carriers of the –735 T allele had lower degenerative grades of LDD [79]. MMP-9 expression and activity are also increased in degenerative discs. The T allele of the –1562 C/T polymorphism increases transcriptional activity of *MMP9* and was associated with a high risk of LDD [80]. Furthermore, a non-synonymous substitution Gln279Arg (rs17576 G/A) in exon 6 of *MMP9* was linked with LDH [41]. This potentially functional SNP may affect substrate binding [41].

MMP-3, also known as stromelysin-1, is involved in the degradation of several collagen types (II, III, IV, IX and X), proteoglycans, fibronectin, elastin and laminin [63]. Increased *MMP3* mRNA levels were observed in ACL fibroblasts compared to medial collateral ligament fibroblasts [81]. Although several studies have reported reduced *MMP3* gene expression in both degenerative and ruptured Achilles tendons [64, 66], de Mos et al. [66] found that MMP-3 activity was actually increased in AT tissues. Three polymorphisms within the *MMP3* gene were investigated for their association with Achilles tendon pathology (ATP), namely rs679620 A/G (Glu45Lys), rs591058 (1547 T/C) and rs650108 (495 A/G). Raleigh et al. [82] noted that the rs679620 GG, rs591058 CC and rs650108 AA genotypes were significantly overrepresented in the AT group. Conversely, the inferred A-T-G haplotype, consisting of the alternate alleles, was underrepresented in the AT group when compared to the controls. In addition, a significant interaction between the *MMP3* rs679620 and the *COL5A1* rs12722 polymorphisms and the risk of AT was observed. No associations were found between the 3 *MMP3* variants and the risk of Achilles tendon rupture [82]. The authors hypothesized that the rs679620 SNP may be in linkage disequilibrium with a functional promoter polymorphism (rs3025058 5A/6A) at position –1612 of the *MMP3* gene. The 5A allele has been shown to have increased transcriptional activity compared to the 6A allele [83] and was also associated with an increased risk of ACL ruptures [84] and intervertebral disc disease [85, 86].

Two other groups of proteinases, related to the MMPs, also play key roles in matrix degradation and remodelling, namely the ADAMs (a disintegrin and metalloproteinase) and ADAMTS (ADAMs with thrombospondin motifs). Recently, 3 intronic polymorphisms (rs151058, rs229052 and rs162502) of the *ADAMTS5* gene were associated with LDD [87]. ADAMTS-5 is a metallopeptidase involved in the cleavage of aggrecan. Increased protein expression has been reported in intervertebral discs [88] whereas decreased mRNA levels were reported in degenerative Achilles tendons [64]. No associations were observed with polymorphisms within genes encoding several ADAMs and ADAMTSs and the risk of ATP in 2 populations, though, interestingly, the *ADAMTS14* rs4747096 GG variant appeared to be linked with a delayed onset of ATP [89].

MMP activity is regulated by the tissue inhibitors of metalloproteinases (TIMPs) which are essential regulators of ECM turnover and remodelling. Increased TIMP expression was reported in tendon and ligament tissue and is suggestive of the tissue's attempt at regeneration and matrix turnover [64, 65]. To date, a genetic variant (rs4789932 C/T) within the promoter region of *TIMP2* has been associated with an increased risk of ATP in a combined Caucasian cohort (South African and Australian participants) [89]. Further investigation is needed to explore genetic markers in the *TIMP1*, *TIMP3* and *TIMP4* genes for their association with musculoskeletal injuries.

Cytokines and Signalling Factors

Cytokines, growth factors and signalling factors play key roles in matrix turnover, inflammation, wound healing, angiogenesis, hypoxia and pain perception amongst others. Moreover, cytokines are upregulated in mechanically loaded tendons and ligaments [90–92]. To date, the genes encoding several of these molecules have been implicated with the risk of musculoskeletal injuries.

The interleukins (ILs) play key roles in cell signalling pathways, upregulating several critical downstream cascades as well as having key roles in inflammation. A potentially functional polymorphism (rs1800587 G/A) [93, 94] in the 5′-UTR of the *IL1A* gene, encoding the α-subunit of IL-1, was associated with LDD [95]. In a pathway analysis, the genes encoding IL-1β *(IL1B)*, IL-1 receptor antagonist *(IL1RN)* and IL-6 *(IL6)* were collectively associated with the risk of AT as part of an inferred allele combination with the previously associated *COL5A1* rs12722 SNP [96]. IL-1β is a potent pro-inflammatory cytokine which activates numerous signal transduction systems. Studies have shown that repetitive loading of tendons results in upregulation of IL-1β levels, which subsequently upregulates expression of itself and other inflammatory mediators, including IL-6, cyclooxygenase-2, prostaglandin E_2, TGF-β and several MMPs [91, 92, 97]. The rs16944 T/C and rs1143627 C/T variants, situated at positions –511 and –31 of the promoter, respectively, were investigated. Both the –511 C allele and the –31 T allele are linked with increased *IL1B* mRNA expression [98]. IL-1ra is the receptor antagonist for IL-1α and IL-1β. An 86-bp variable nucleotide tandem repeat (rs2234663) in intron 2 of the *IL1RN* gene contains several potential protein-binding sites [99]. However, there is conflicting evidence regarding the functionality of this polymorphism with some studies indicating the A2 repeat allele is associated with increased IL-1ra expression and others the reverse [99–101]. IL-6, encoded by *IL6*, is an inflammatory cytokine believed to play a key role in apoptosis, and expression levels of this cytokine can be correlated with severity of inflammation [102]. Increased IL-6 expression has been noted in tendon pathology [103, 104], after cyclic stretching of human tendon fibroblasts [102] and also in ACL-deficient knees [105]. The rs1800795 G/C polymorphism is situated at position –172 of the promoter, with the G allele associated with increased transcription [106]. Recently, this SNP has been associated with a risk of intervertebral disc disease, both independently and as a haplotype [44]. IL-6 signalling is initiated upon binding of the IL to its receptor, encoded by the *IL6R* gene. A non-synonymous SNP within *IL6R* (rs2228145 A/C; Asp358Ala) was independently associated with a reduced risk of CTS [107]. Furthermore, interactions between the *IL6R* rs2228145 polymorphism and the previously associated *COL5A1* [22] and *BGN* [58] variants were observed to modulate the risk of CTS [107].

IL-10 is an anti-inflammatory cytokine with pleiotropic effects in immunoregulation and inflammation, and the ability to inhibit the synthesis of IL-1, IL-6, tumour necrosis factor-α and TGF-β [108]. Several polymorphisms within the promoter region of *IL10* may affect transcriptional regulation of the gene. The –1082 A/G (rs1800896) and –592 A/C (rs1800872) polymorphisms were associated with LDD, with the –1082 AA and –592 AA genotypes conferring an increased risk of LDD [109]. The –592 CC genotype is linked with increased *IL10* mRNA expression compared to the A allele [109, 110] but there is conflicting evidence regarding which –1082 allele results in the increased transcriptional activity of the gene [109].

IL-18 is a pro-inflammatory cytokine belonging to the IL-1 superfamily. IL-18 binds to the IL-18 receptor 1 (IL-18R1) and triggers the recruitment of the IL-18 receptor accessory protein (IL-18RAP) which then initiates IL-18 signalling [111]. *IL18R1* and *IL18RAP* are both localized on chromosome 2q12.1. A potentially functional variant (rs2287037) in the *IL18R1* promoter encodes an A-to-G transition and may alter a transcription factor binding site [111]. There was a trend for this polymorphism to be associated with LDD, with the CC genotype overrepresented in cases compared to the controls [112]. Disc desiccation is considered to be an earlier stage of the degenerative process and is measured by disc signal intensity as a quantitative measure of disc degeneration. The A allele of the functional rs917997 A/G variant located in the *IL18RAP* 3′-UTR was associated with lower disc signal intensity in participants with LDD [18]. The A allele has previously been linked with reduced *IL-18RAP* mRNA expression in whole-blood samples [113]. Additionally, the *IL18RAP* promoter rs1420106 T/C and intronic rs1420100 T/G polymorphisms were associated with increased disc degeneration in participants with LDD [18, 112, 114].

Angiogenesis is the formation of new blood vessels from pre-existing vasculature and is considered to play an important role in ECM remodelling. Increased levels of angiogenic cytokines have been reported in both ruptured tendons and ligaments as well as in degenerative tendons [115–117]. Vascular endothelial growth factor A (VEGF-A) is a critical regulator of angiogenesis which can be induced by both biological stimuli and mechanical loading [118, 119]. VEGF activity is mediated through its receptor kinase insert domain receptor, encoded by the *KDR* gene. Functional polymorphisms within *VEGFA* (rs699947 –2578 C/A, rs1570360 –1154 G/A and rs2010963 –634 G/C) and *KDR* (rs2071559 G/A and rs1870377 T/A) have recently been implicated with the risk of ACL ruptures [120]. The –2578 C allele, –1154 G allele and –634 C allele have all been linked with an increased VEGF expression [121, 122]. The –2578 CC genotype and C allele were associated with an increased risk of non-contact ACL ruptures and the –1154 GA genotype with a reduced risk of rupture in the control participants [120]. Interestingly, the inferred C-G-C haplotype (–2578 C/A, –1154 G/A, –634 G/C) consisting of the high-VEGF-producing alleles was linked with an increased risk of non-contact ACL ruptures in females. Conversely, the A-A-G haplotype was associated with a decreased risk of non-contact ACL ruptures in all participants, and this haplotype correlates with reduced circulating VEGF levels in vivo and decreased *VEGF* transcription [122]. It is worth mentioning that VEGF expression may stimulate MMP expression levels [123] which could negatively impact ECM homeostasis.

The *KDR* rs2071559 SNP involves a G-to-A transition at position –604 of the *KDR* promoter and the rs1870377 SNP a non-synonymous substitution in exon 11 (+1719 T/A; His472Gln). Both minor alleles potentially affect VEGF signalling: the –604 G allele alters a transcription factor binding site resulting in decreased transcription, and the +1719 A allele affects the efficacy of VEGF binding to KDR [124]. In the present study, the –604 GA genotype was associated with a 2.2-fold decreased risk of ACL ruptures in female participants. Haplotype analysis further implicated the gene, as the G-A (–604 G/A, +1719 T/A) inferred haplotype was associated with an increased risk of ACL ruptures in all participants, and in males and females separately [120].

Angiogenesis may also be initiated under hypoxic conditions by hypoxia-inducible factor-1α (HIF-1α) which can induce transcription of the *VEGF* gene. Furthermore, HIF-1α and VEGF levels were both shown to be increased after cyclic straining of tendon fibroblasts [119]. HIF-1α regulates the cellular response to hypoxia but can also be induced under normoxic conditions [125] such as mechanical loading. A nonsynonymous SNP +1790 G/A (rs11549467) in exon 12 of the *HIF1A* gene, representing an alanine-to-threonine substitution at codon 588, was associated with LDD [126]. The 1790 A allele was previously linked with increased transcriptional activity of the *HIF1A* gene [127].

Growth differentiation factor-5 (GDF-5), a member of the TGF-β superfamily, is a regulator of cell growth and differentiation with an essential role in skeletal and joint development [128, 129]. Achilles tendons in mutant Gdf5-deficient mice were found to be significantly weaker with 40% less collagen content compared to healthy tendons [130]. A functional polymorphism (rs143383), situated in the 5′-UTR, involves a T-to-C transition, the T allele correlating with decreased *GDF5* expression [131]. Furthermore, the TT genotype was associated with the risk of developing ATP [132], and the T allele with the risk of LDD [133], but no associations were noted with the risk of ACL rupture [134].

Vitamin D receptor (VDR) is a transcription factor primarily involved in bone mineralization and remodelling but also regulates several key metabolic and immune pathways. The functional *Fok*I polymorphism (rs2228570) represents a C-to-T transition in the *VDR* start codon (Met1Thr). Consequently, the T allele results in an alternate start codon, which forms a longer (427 aa compared to 424 aa) and less transcriptionally potent protein [135]. The Ff and ff genotypes (CT and TT, respectively) were associated with more disc degeneration compared to carriers of the FF (CC) genotype [136]. The *Taq*I RFLP (rs731236 T/C) encodes a synonymous substitution (Ile352Ile) in exon 9 of *VDR*. The *Taq*I Tt and tt genotypes (TC and CC, respectively) were associated with disc degeneration in several populations [54, 136–139]. However, as this variant is not functional, it is hypothesized that it may be in linkage with another disease-predisposing functional variant [140].

Apoptosis

Tendinopathic tendons have been characterized by excessive levels of apoptotic cells, specifically in patellar tendinosis [141], rotator cuff tendons [104, 142] and after mechanical loading of the tibialis anterior tendon in a rodent model [143]. Millar et al. [104] hypothesized that mechanical overloading may induce prolonged activity of cytokines and apoptotic mediators resulting in excessive apoptosis and a subsequently weakened collagenous matrix with greater degradation. Apoptosis may be initiated through the intrinsic and extrinsic pathways. To clarify, the intrinsic pathway is mediated via mitochondria and the extrinsic pathway through the binding of extracellular death ligands to cell surface death receptors (forming a death-inducing signalling complex). Apoptotic cell death is mediated by the activation of caspases (cysteine-containing aspartate proteinases) – a group of proteolytic enzymes also involved in inflammation. Caspase-8 and caspase-9 are both initiator caspases responsible for the activation of caspase-3, an effector caspase, via proteolytic cleavage. Caspase-3 then cleaves other cellular targets and caspases, thereby triggering apoptosis.

Caspase-8, encoded by *CASP8*, functions via the extrinsic apoptosis pathway and is upregulated by IL-6. Two functional polymorphisms within the *CASP8* gene – rs3834129 ins/del and rs1045485 G/C – were implicated with AT in males [144]. The rs3834129 polymorphism encodes a 6-nucleotide deletion variant in the *CASP8* promoter at position –652, and the del allele abolishes an Sp1 binding element resulting in decreased *CASP8* transcription [145]. The non-synonymous rs1045485 SNP, in exon 9, results in a residue change from aspartate to histidine at codon 302 and is considered to affect caspase-8 interactions [145]. Degeneration in intervertebral discs is thought to occur primarily through the intrinsic pathway of apoptosis, which is mediated by caspase-9 [146–148]. Increased caspase-9 expression was reported in degenerative discs compared to normal disc tissue [146, 147] and a non-synonymous SNP (rs1052576 G/A; Gln221Arg) in the coding region of *CASP9* has been implicated with the risk of LDD [149]. Specifically, carriers of the AA genotype were found to have a 1.91-fold increased risk of LDH and greater disc degeneration [149]. The Gln221Arg substitution potentially alters protein conformation and potentially alters protein-protein interactions and protein function [150].

As described above, the extrinsic pathway of apoptosis is initiated by the formation of a death-inducing signalling complex such as the cross-linking of the Fas ligand (FasL) to its receptor. FasL is a member of the tumour necrosis factor superfamily, and the Fas cell surface death receptor is a transmembrane receptor belonging to the tumour necrosis factor receptor superfamily. Increased expressions of both the ligand and receptor were reported in herniated lumbar discs [151, 152] and functional polymorphisms within the *FASL* and *FAS* genes associated with LDD [153]. The *FASL* –844 C/T (rs763110) SNP is located in a putative binding motif for the CAAT/enhancer-binding protein-β element. The –844 C allele showed over double the activity than the –844 T allele and was also correlated with increased *FASL* expression [154]. The rs2234767 promoter polymorphism in the *FAS* gene, coding for the receptor, represents a G-to-A transition at position –1377. The A allele destroys an Sp1 binding site, thereby resulting in reduced transcriptional activity and *FAS* expression [155]. The *FASL* –844 CT and CC genotypes and the *FAS* –1377 GA and AA genotypes were associated with LDD [153]. Furthermore, the said genotypes were also linked with higher grades of disc degeneration compared to their homozygous counterparts (–844 TT and –1377 GG, respectively) [153]. B-cell lymphoma-2, encoded by the *BCL2* gene, is a regulator of apoptosis and is considered anti-apoptotic. A functional polymorphism (rs2279115; –938 C/A) identified in the promoter region was linked with LDD, with the CA and AA genotypes, as well as the A allele, conferring an increased risk compared to the CC genotype and C allele, respectively [156].

Concluding Remarks

To date, several genetic loci implicating an array of genes encoding proteins with diverse functions have been implicated in the susceptibility to musculoskeletal injuries. The large majority of these studies have followed a case-control genetic association approach, which is a limitation. Only a small proportion of these associations have been replicated in independent populations having the same or similar phenotype [157]. There is most likely evidence supporting these implicated regions as candidate genes partially contributing to the susceptibility to musculoskeletal injuries. However, there is limited functional evidence to support some of the genomic intervals and specific polymorphisms implicated in these associations. Evidence for a functional hypothesis is emerging for the promoter region of *COL1A1* [13] and the 3′-UTR of the *COL5A1* gene [24]. Regulating homeostasis within the ECM of the tissues implicated in these musculoskeletal injuries is complex, and therefore it is not surprising that a myriad of networks and pathways have been implicated thus far.

It has been proposed that genetic susceptibility may contribute to the interindividual variation in the response to mechanical loading, which may consequently affect an individual's tissue integrity and biomechanical function, healing capacity and adaptive responses to load [158]. It is critical that we start understanding the functional role of these genomic signatures underpinning the genetic boundaries implicated in the associations described in table 1. There is, therefore, a need to explore the functional consequence of the individual loci. However, unravelling the net effect of all these loci in response to mechanical load is required if we want to understand the clinical relevance. One of the steps towards this functional end is exploring the genomic signatures underpinning these loci and then sequence mining these intervals and boundaries more carefully to identify critical regions of biological relevance and exploring these signatures for susceptibility across populations. Such studies will inform us of the biologically significant 'susceptibility genomics map' for musculoskeletal injuries. The next step would be to explore the functional map underpinning these susceptibilities in an attempt to unravel the complexity of understanding the susceptibility to injury and response to load. Moreover, we have not begun to unravel the effects of epigenetic regulation on risk susceptibility [159–161].

Although our current understanding of the pathobiology of these diverse phenotypes is still in its infancy, we have made much progress in understanding that injury susceptibility is complex and is most likely influenced by several combinations of genomic signatures, interconnecting diverse biological pathways to regulate the effective responses to load. Research in these areas could be accelerated if international consortia were established.

References

1 Ljungqvist A, Schwellnus MP, Bachl N, Collins M, Cook J, Khan KM, et al: International Olympic Committee consensus statement: molecular basis of connective tissue and muscle injuries in sport. Clin Sports Med 2008;27:231–239.

2 Posthumus M, Collins M, September AV, Schwellnus MP: The intrinsic risk factors for ACL ruptures: an evidence-based review. Phys Sportsmed 2011;39: 62–73.

3 Mann V, Hobson EE, Li B, Stewart TL, Grant SF, Robins SP, et al: A COL1A1 Sp1 binding site polymorphism predisposes to osteoporotic fracture by affecting bone density and quality. J Clin Invest 2001; 107:899–907.

4 Stępień-Słodkowska M, Ficek K, Eider J, Leońska-Duniec A, MacIejewska-Karłowska A, Sawczuk M, et al: The +1245G/T polymorphisms in the collagen type I alpha 1 *(COL1A1)* gene in Polish skiers with anterior cruciate ligament injury. Biol Sport 2013;30: 57–60.

5 Khoschnau S, Melhus H, Jacobson A, Rahme H, Bengtsson H, Ribom E, et al: Type I collagen alpha1 Sp1 polymorphism and the risk of cruciate ligament ruptures or shoulder dislocations. Am J Sports Med 2008;36:2432–2436.

6 Posthumus M, September AV, Keegan M, O'Cuinneagain D, Van der Merwe W, Schwellnus MP, et al: Genetic risk factors for anterior cruciate ligament ruptures: *COL1A1* gene variant. Br J Sports Med 2009;43:352–356.

7 Collins M, Posthumus M, Schwellnus MP: The *COL1A1* gene and acute soft tissue ruptures. Br J Sports Med 2010;44:1063–1964.

8 Ficek K, Cieszczyk P, Kaczmarczyk M, Maciejewska-Karłowska A, Sawczuk M, Cholewinski J, et al: Gene variants within the *COL1A1* gene are associated with reduced anterior cruciate ligament injury in professional soccer players. J Sci Med Sport 2013;16:396–400.

9 Posthumus M, September AV, Schwellnus MP, Collins M: Investigation of the Sp1-binding site polymorphism within the *COL1A1* gene in participants with Achilles tendon injuries and controls. J Sci Med Sport 2009;12:184–189.

10 Bere T, Flørenes TW, Krosshaug T, Koga H, Nordsletten L, Irving C, et al: Mechanisms of anterior cruciate ligament injury in World Cup alpine skiing: a systematic video analysis of 20 cases. Am J Sports Med 2011;39:1421–1429.

11 Pluijm SMF, van Essen HW, Bravenboer N, Uitterlinden AG, Smit JH, Pols HAP, et al: Collagen type I alpha1 Sp1 polymorphism, osteoporosis, and intervertebral disc degeneration in older men and women. Ann Rheum Dis 2004;63:71–77.

12 Tilkeridis C, Bei T, Garantziotis S, Stratakis CA: Association of a *COL1A1* polymorphism with lumbar disc disease in young military recruits. J Med Genet 2005;42:e44.

13 Jin H, van't Hof RJ, Albagha OME, Ralston SH: Promoter and intron 1 polymorphisms of *COL1A1* interact to regulate transcription and susceptibility to osteoporosis. Hum Mol Genet 2009;18:2729–2738.

14 Ireland D, Harrall R, Curry V, Holloway G, Hackney R, Hazleman B, et al: Multiple changes in gene expression in chronic human Achilles tendinopathy. Matrix Biol 2001;20:159–169.

15 Kluivers KB, Dijkstra JR, Hendriks JCM, Lince SL, Vierhout ME, van Kempen LCL: *COL3A1* 2209G>A is a predictor of pelvic organ prolapse. Int Urogynecol J Pelvic Floor Dysfunct 2009;20:1113–1118.

16 O'Connell K, Knight H, Ficek K, Leonska-Duniec A, Maciejewska-Karlowska A, Sawczuk M, et al: Interactions between collagen gene variants and risk of anterior cruciate ligament rupture. Eur J Sport Sci 2014;29:1–10.

17 Stępień-Słodkowska M, Ficek K, Maciejewska-Karłowska A, Sawczuk M, Ziętek P, Król P, et al: Overrepresentation of the *COL3A1* AA genotype in Polish skiers with anterior cruciate ligament injury. Biol Sport 2014;32:143–147.

18 Videman T, Saarela J, Kaprio J, Näkki A, Levälahti E, Gill K, et al: Associations of 25 structural, degradative, and inflammatory candidate genes with lumbar disc desiccation, bulging, and height narrowing. Arthritis Rheum 2009;60:470–481.

19 Mokone GG, Schwellnus MP, Noakes TD, Collins M: The *COL5A1* gene and Achilles tendon pathology. Scand J Med Sci Sports 2006;16:19–26.

20 September AV, Cook J, Handley CJ, van der Merwe L, Schwellnus MP, Collins M: Variants within the *COL5A1* gene are associated with Achilles tendinopathy in two populations. Br J Sports Med 2009;43: 357–365.

21 Posthumus M, September AV, O'Cuinneagain D, van der Merwe W, Schwellnus MP, Collins M: The *COL5A1* gene is associated with increased risk of anterior cruciate ligament ruptures in female participants. Am J Sports Med 2009;37:2234–2240.

22 Burger M, de Wet H, Collins M: The *COL5A1* gene is associated with increased risk of carpal tunnel syndrome. Clin Rheumatol 2015;34:767–774.

23 Abrahams Y, Laguette M-J, Prince S, Collins M: Polymorphisms within the *COL5A1* 3′-UTR that alters mRNA structure and the *MIR608* gene are associated with Achilles tendinopathy. Ann Hum Genet 2013;77:204–214.

24 Laguette M-J, Abrahams Y, Prince S, Collins M: Sequence variants within the 3′-UTR of the *COL5A1* gene alters mRNA stability: implications for musculoskeletal soft tissue injuries. Matrix Biol 2011;30: 338–345.
25 Collins M, Posthumus M: Type V collagen genotype and exercise-related phenotype relationships. Exerc Sport Sci Rev 2011;39:4:191–198.
26 Annunen S, Paassilta P, Lohiniva J, Perälä M, Pihlajamaa T, Karppinen J, et al: An allele of *COL9A2* associated with intervertebral disc disease. Science 1999;285:409–412.
27 Jim JJT, Noponen-Hietala N, Cheung KMC, Ott J, Karppinen J, Sahraravand A, et al: The TRP2 allele of *COL9A2* is an age-dependent risk factor for the development and severity of intervertebral disc degeneration. Spine 2005;30:2735–2742.
28 Higashino K, Matsui Y, Yagi S, Takata Y, Goto T, Sakai T, et al: The alpha2 type IX collagen tryptophan polymorphism is associated with the severity of disc degeneration in younger patients with herniated nucleus pulposus of the lumbar spine. Int Orthop 2007;31:107–111.
29 Kales SN, Linos A, Chatzis C, Sai Y, Halla M, Nasioulas G, et al: The role of collagen IX tryptophan polymorphisms in symptomatic intervertebral disc disease in Southern European patients. Spine 2004;29: 1266–1270.
30 Seki S, Kawaguchi Y, Mori M, Mio F, Chiba K, Mikami Y, et al: Association study of *COL9A2* with lumbar disc disease in the Japanese population. J Hum Genet 2006;51:1063–1067.
31 Karppinen J, Pääkkö E, Räinä S, Tervonen O, Kurunlahti M, Nieminen P, et al: Magnetic resonance imaging findings in relation to the *COL9A2* tryptophan allele among patients with sciatica. Spine 2002;27: 78–83.
32 Paassilta P, Lohiniva J, Göring HH, Perälä M, Räinä SS, Karppinen J, et al: Identification of a novel common genetic risk factor for lumbar disk disease. JAMA 2001;285:1843–1849.
33 Karppinen J, Pääkkö E, Paassilta P, Lohiniva J, Kurunlahti M, Tervonen O, et al: Radiologic phenotypes in lumbar MR imaging for a gene defect in the *COL9A3* gene of type IX collagen. Radiology 2003; 227:143–148.
34 Fichard A, Kleman JP, Ruggiero F: Another look at collagen V and XI molecules. Matrix Biol 1995;14: 515–531.
35 Wenstrup RJ, Smith SM, Florer JB, Zhang G, Beason DP, Seegmiller RE, et al: Regulation of collagen fibril nucleation and initial fibril assembly involves coordinate interactions with collagens V and XI in developing tendon. J Biol Chem 2011;286:20455–20465.
36 Mio F, Chiba K, Hirose Y, Kawaguchi Y, Mikami Y, Oya T, et al: A functional polymorphism in *COL11A1,* which encodes the alpha 1 chain of type XI collagen, is associated with susceptibility to lumbar disc herniation. Am J Hum Genet 2007;81:1271–1277.
37 Hay M, Patricios J, Collins R, Branfield A, Cook J, Handley CJ, et al: Association of type XI collagen genes with chronic Achilles tendinopathy in independent populations from South Africa and Australia. Br J Sports Med 2013;47:569–574.
38 Posthumus M, September AV, O'Cuinneagain D, van der Merwe W, Schwellnus MP, Collins M: The association between the *COL12A1* gene and anterior cruciate ligament ruptures. Br J Sports Med 2010;44: 1160–1165.
39 Mokone GG, Gajjar M, September AV, Schwellnus MP, Greenberg J, Noakes TD, et al: The guanine-thymine dinucleotide repeat polymorphism within the tenascin-C gene is associated with Achilles tendon injuries. Am J Sports Med 2005;33:1016–1021.
40 Saunders CJ, van der Merwe L, Posthumus M, Cook J, Handley CJ, Collins M, et al: Investigation of variants within the *COL27A1* and *TNC* genes and Achilles tendinopathy in two populations. J Orthop Res 2013;31:632–637.
41 Hirose Y, Chiba K, Karasugi T, Nakajima M, Kawaguchi Y, Mikami Y, et al: A functional polymorphism in *THBS2* that affects alternative splicing and MMP binding is associated with lumbar-disc herniation. Am J Hum Genet 2008;82:1122–1129.
42 El Khoury L, Posthumus M, Collins M, van der Merwe W, Handley C, Cook J, et al: *ELN* and *FBN2* gene variants as risk factors for two sports-related musculoskeletal injuries. Int J Sports Med 2015;36:333–337.
43 Seki S, Kawaguchi Y, Chiba K, Mikami Y, Kizawa H, Oya T, et al: A functional SNP in *CILP,* encoding cartilage intermediate layer protein, is associated with susceptibility to lumbar disc disease. Nat Genet 2005;37:607–612.
44 Kelempisioti A, Eskola PJ, Okuloff A, Karjalainen U, Takatalo J, Daavittila I, et al: Genetic susceptibility of intervertebral disc degeneration among young Finnish adults. BMC Med Genet 2011;12:153.
45 Yanagishita M: Function of proteoglycans in the extracellular matrix. Acta Pathol Jpn 1993;43:283–293.
46 Nastase MV, Young MF, Schaefer L: Biglycan: a multivalent proteoglycan providing structure and signals. J Histochem Cytochem 2012;60:963–975.
47 Corps AN, Robinson AHN, Movin T, Costa ML, Hazleman BL, Riley GP: Increased expression of aggrecan and biglycan mRNA in Achilles tendinopathy. Rheumatology 2006;45:291–294.

48 Corps AN, Robinson AHN, Movin T, Costa ML, Ireland DC, Hazleman BL, et al: Versican splice variant messenger RNA expression in normal human Achilles tendon and tendinopathies. Rheumatology 2004; 43:969–972.
49 Young K, Samiric T, Feller J, Cook J: Extracellular matrix content of ruptured anterior cruciate ligament tissue. Knee 2011;18:242–246.
50 Johnson JS, Morscher MA, Jones KC, Moen SM, Klonk CJ, Jacquet R, Landis WJ: Gene expression differences between ruptured anterior cruciate ligaments in young male and female subjects. J Bone Joint Surg Am 2015;97:71–79.
51 Doege KJ: A human-specific polymorphism in the coding region of the aggrecan gene. Variable number of tandem repeats produces a range of core protein sizes in the general population. J Biol Chem 1997; 272:13974–13979.
52 Kawaguchi Y, Osada R, Kanamori M, Ishihara H, Ohmori K, Matsui H, et al: Association between an aggrecan gene polymorphism and lumbar disc degeneration. Spine 1999;24:2456–2460.
53 Cong L, Pang H, Xuan D, Tu GJ: Association between the expression of aggrecan and the distribution of aggrecan gene variable number of tandem repeats with symptomatic lumbar disc herniation in Chinese Han of Northern China. Spine 2010;35: 1371–1376.
54 Eser B, Cora T, Eser O, Kalkan E, Haktanir A, Erdogan MO, et al: Association of the polymorphisms of vitamin D receptor and aggrecan genes with degenerative disc disease. Genet Test Mol Biomarkers 2010;14:313–317.
55 Mashayekhi F, Shafiee G, Kazemi M, Dolati P: Lumbar disk degeneration disease and aggrecan gene polymorphism in northern Iran. Biochem Genet 2010;48:684–689.
56 Eser O, Eser B, Cosar M, Erdogan MO, Aslan A, Yıldız H, et al: Short aggrecan gene repetitive alleles associated with lumbar degenerative disc disease in Turkish patients. Genet Mol Res 2011;10:1923–1930.
57 Mannion S, Mtintsilana A, Posthumus M, van der Merwe W, Hobbs H, Collins M, et al: Genes encoding proteoglycans are associated with the risk of anterior cruciate ligament ruptures. Br J Sports Med 2014;22:1640–1646.
58 Burger MC, de Wet H, Collins M: The *BGN* and *ACAN* genes and carpal tunnel syndrome. Gene 2014;551:160–166.
59 Hildebrand A, Romarís M, Rasmussen LM, Heinegård D, Twardzik DR, Border WA, et al: Interaction of the small interstitial proteoglycans biglycan, decorin and fibromodulin with transforming growth factor beta. Biochem J 1994;302:527–534.
60 Lorenzo P, Aspberg A, Önnerfjord P, Bayliss MT, Neame PJ, Heinegård D: Identification and characterization of asporin. A novel member of the leucine-rich repeat protein family closely related to decorin and biglycan. J Biol Chem 2001;276:12201–12211.
61 Song Y-Q, Cheung KMC, Ho DWH, Poon SCS, Chiba K, Kawaguchi Y, et al: Association of the asporin D14 allele with lumbar-disc degeneration in Asians. Am J Hum Genet 2008;82:744–747.
62 Somerville RPT, Oblander SA, Apte SS: Matrix metalloproteinases: old dogs with new tricks. Genome Biol 2003;4:216.
63 Pasternak B, Aspenberg P: Metalloproteinases and their inhibitors – diagnostic and therapeutic opportunities in orthopedics. Acta Orthop 2009;80:693–703.
64 Jones GC, Corps AN, Pennington CJ, Clark IM, Edwards DR, Bradley MM, et al: Expression profiling of metalloproteinases and tissue inhibitors of metalloproteinases in normal and degenerate human Achilles tendon. Arthritis Rheum 2006;54:832–842.
65 Tang Z, Yang L, Xue R, Zhang J, Wang Y, Chen PC, et al: Differential expression of matrix metalloproteinases and tissue inhibitors of metalloproteinases in anterior cruciate ligament and medial collateral ligament fibroblasts after a mechanical injury: involvement of the p65 subunit of NF-kappaB. Wound Repair Regen 2009;17:709–716.
66 De Mos M, van El B, DeGroot J, Jahr H, van Schie HTM, van Arkel ER, et al: Achilles tendinosis: changes in biochemical composition and collagen turnover rate. Am J Sports Med 2007;35:1549–1556.
67 Posthumus M, Collins M, van der Merwe L, O'Cuinneagain D, van der Merwe W, Ribbans WJ, et al: Matrix metalloproteinase genes on chromosome 11q22 and the risk of anterior cruciate ligament (ACL) rupture. Scand J Med Sci Sport 2012;16;22: 523–533.
68 Song Y-Q, Ho DWH, Karppinen J, Kao PYP, Fan B-J, Luk KDK, et al: Association between promoter –1607 polymorphism of *MMP1* and lumbar disc disease in Southern Chinese. BMC Med Genet 2008;9: 38.
69 Godoy-Santos A, Cunha M V, Ortiz RT, Fernandes TD, Mattar R, dos Santos MCLG: MMP-1 promoter polymorphism is associated with primary tendinopathy of the posterior tibial tendon. J Orthop Res 2013;31:1103–1107.
70 Rutter JL, Mitchell TI, Butticè G, Meyers J, Gusella JF, Ozelius LJ, et al: A single nucleotide polymorphism in the matrix metalloproteinase-1 promoter creates an Ets binding site and augments transcription. Cancer Res 1998;58:5321–5325.

71 Coon CI, Fiering S, Gaudet J, Wyatt CA, Brinckerhoff CE: Site controlled transgenic mice validating increased expression from human matrix metalloproteinase (MMP-1) promoter due to a naturally occurring SNP. Matrix Biol 2009;28:425–431.
72 Baroneza JE, Godoy-Santos A, Ferreira Massa B, Boçon de Araujo Munhoz F, Diniz Fernandes T, Leme Godoy dos Santos MC: MMP-1 promoter genotype and haplotype association with posterior tibial tendinopathy. Gene 2014;547:334–337.
73 Godoy-Santos A, Ortiz RT, Junior RM, Fernandes TD, Santos MCLG: MMP-8 polymorphism is genetic marker to tendinopathy primary posterior tibial tendon. Scand J Med Sci Sport 2014;24:220–223.
74 Wang H, Parry S, Macones G, Sammel MD, Ferrand PE, Kuivaniemi H, et al: Functionally significant SNP MMP8 promoter haplotypes and preterm premature rupture of membranes (PPROM). Hum Mol Genet 2004;13:2659–2669.
75 Crean JK, Roberts S, Jaffray DC, Eisenstein SM, Duance VC: Matrix metalloproteinases in the human intervertebral disc: role in disc degeneration and scoliosis. Spine 1997;22:2877–2884.
76 Roberts S, Caterson B, Menage J, Evans EH, Jaffray DC, Eisenstein SM: Matrix metalloproteinases and aggrecanase: their role in disorders of the human intervertebral disc. Spine 2000;25:3005–3013.
77 Dong DM, Yao M, Liu B, Sun CY, Jiang YQ, Wang YS: Association between the –1306C/T polymorphism of matrix metalloproteinase-2 gene and lumbar disc disease in Chinese young adults. Eur Spine J 2007;16:1958–1961.
78 Yu C, Zhou Y, Miao X, Xiong P, Tan W, Lin D: Functional haplotypes in the promoter of matrix metalloproteinase-2 predict risk of the occurrence and metastasis of esophageal cancer. Cancer Res 2004;64:7622–7628.
79 Zhang Y, Gu Z, Qiu G: Association of the polymorphism of *MMP2* with the risk and severity of lumbar disc degeneration in the Chinese Han population. Eur Rev Med Pharmacol Sci 2013;17:1830–1834.
80 Sun Z, Miao L, Zhang Y, Ming L: Association between the –1562 C/T polymorphism of matrix metalloproteinase-9 gene and lumbar disc disease in the young adult population in North China. Connect Tissue Res 2009;50:181–185.
81 Zhang J, Yang L, Tang Z, Xue R, Wang Y, Luo Z, et al: Expression of MMPs and TIMPs family in human ACL and MCL fibroblasts. Connect Tissue Res 2009;50:7–13.
82 Raleigh SM, van der Merwe L, Ribbans WJ, Smith RKW, Schwellnus MP, Collins M: Variants within the *MMP3* gene are associated with Achilles tendinopathy: possible interaction with the *COL5A1* gene. Br J Sports Med 2009;43:514–520.
83 Ye S, Eriksson P, Hamsten A, Kurkinen M, Humphries SE, Henney AM: Progression of coronary atherosclerosis is associated with a common genetic variant of the human stromelysin-1 promoter which results in reduced gene expression. J Biol Chem 1996;271:13055–13060.
84 Malila S, Yuktanandana P, Saowaprut S, Jiamjarasrangsi W, Honsawek S: Association between matrix metalloproteinase-3 polymorphism and anterior cruciate ligament ruptures. Genet Mol Res 2011;10:4158–4165.
85 Takahashi M, Haro H, Wakabayashi Y, Komori H, Shinomiya K: Intervertebral disc with 5A/6A polymorphism in the promoter of the human matrix metalloproteinase-3 gene. J Bone Joint Surg Br 2001;83:491–495.
86 Zawilla NH, Darweesh H, Mansour N, Helal S, Taha FM, Awadallah M, et al: Matrix metalloproteinase-3, vitamin D receptor gene polymorphisms, and occupational risk factors in lumbar disc degeneration. J Occup Rehabil 2014;24:370–381.
87 Wu N, Chen J, Liu H, Zhao L, Liu S, Liu J, et al: The involvement of ADAMTS-5 genetic polymorphisms in predisposition and diffusion tensor imaging alterations of lumbar disc degeneration. J Orthop Res 2014;32:686–694.
88 Zhao CQ, Zhang YH, Jiang SD, Li H, Jiang LS, Dai LY: ADAMTS-5 and intervertebral disc degeneration: the results of tissue immunohistochemistry and in vitro cell culture. J Orthop Res 2011;29:718–725.
89 El Khoury L, Posthumus M, Collins M, Handley CJ, Cook J, Raleigh SM: Polymorphic variation within the *ADAMTS2, ADAMTS14, ADAMTS5, ADAM12* and *TIMP2* genes and the risk of Achilles tendon pathology: a genetic association study. J Sci Med Sport 2013;16:493–498.
90 Evans CH: Cytokines and the role they play in the healing of ligaments and tendons. Sports Med 1999;28:71–76.
91 Yang G, Im H-J, Wang JH-C: Repetitive mechanical stretching modulates IL-1β induced COX-2, MMP-1 expression, and PGE_2 production in human patellar tendon fibroblasts. Gene 2005;363:166–172.
92 Tsuzaki M, Guyton G, Garrett W, Archambault JM, Herzog W, Almekinders L, et al: IL-1B induces COX2, MMP-1, -3 and -13, ADAMTS-4, IL-1B and IL-6 in human tendon cells. J Orthop Res 2003;21:256–264.
93 Eskola PJ, Kjaer P, Sorensen JS, Okuloff A, Wedderkopp N, Daavittila I, et al: Gender difference in genetic association between *IL1A* variant and early lumbar disc degeneration: a three-year follow-up. Int J Mol Epidemiol Genet 2012;3:195–204.

94 Kawaguchi Y, Tochimoto A, Hara M, Kawamoto M, Sugiura T, Saito S, et al: Contribution of single nucleotide polymorphisms of the *IL1A* gene to the cleavage of precursor IL-1alpha and its transcription activity. Immunogenetics 2007;59:441–448.

95 Eskola PJ, Kjaer P, Daavittila IM, Solovieva S, Okuloff A, Sorensen JS, et al: Genetic risk factors of disc degeneration among 12–14-year-old Danish children: a population study. Int J Mol Epidemiol Genet 2010;1:158–165.

96 September AV, Nell E-M, O'Connell K, Cook J, Handley CJ, van der Merwe L, et al: A pathway-based approach investigating the genes encoding interleukin-1β, interleukin-6 and the interleukin-1 receptor antagonist provides new insight into the genetic susceptibility of Achilles tendinopathy. Br J Sports Med 2011;45:13:1040–1047.

97 Thampatty BP, Li H, Im H-J, Wang JH-C: EP4 receptor regulates collagen type-I, MMP-1, and MMP-3 gene expression in human tendon fibroblasts in response to IL-1β treatment. Gene 2007; 386:154–161.

98 Landvik NE, Hart K, Skaug V, Stangeland LB, Haugen A, Zienolddiny S: A specific interleukin-1B haplotype correlates with high levels of *IL1B* mRNA in the lung and increased risk of non-small cell lung cancer. Carcinogenesis 2009;30:1186–1192.

99 Danis VA, Millington M, Hyland VJ, Grennan D: Cytokine production by normal human monocytes: inter-subject variation and relationship to an IL-1 receptor antagonist (IL-1Ra) gene polymorphism. Clin Exp Immunol 1995;99:303–310.

100 Mandrup-Poulsen T, Pociot F, Mølvig J, Shapiro L, Nilsson P, Emdal T, et al: Monokine antagonism is reduced in patients with IDDM. Diabetes 1994;43: 1242–1247.

101 Clay FE, Tarlow JK, Cork MJ, Cox A, Nicklin MJ, Duff GW: Novel interleukin-1 receptor antagonist exon polymorphisms and their use in allele-specific mRNA assessment. Hum Genet 1996;97:723–726.

102 Skutek M, van Griensven M, Zeichen J, Brauer N, Bosch U: Cyclic mechanical stretching enhances secretion of interleukin 6 in human tendon fibroblasts. Knee Surg Sports Traumatol Arthrosc 2001; 9:322–326.

103 Legerlotz K, Jones ER, Screen HRC, Riley GP: Increased expression of IL-6 family members in tendon pathology. Rheumatology 2012;51:1161–1165.

104 Millar NL, Wei AQ, Molloy TJ, Bonar F, Murrell GA: Cytokines and apoptosis in supraspinatus tendinopathy. J Bone Joint Surg Br 2009;91:417–424.

105 Cameron ML, Fu FH, Paessler HH, Schneider M, Evans CH: Synovial fluid cytokine concentrations as possible prognostic indicators in the ACL-deficient knee. Knee Surg Sports Traumatol Arthrosc 1994;2:38–44.

106 Fishman D, Faulds G, Jeffery R, Mohamed-Ali V, Yudkin JS, Humphries S, et al: The effect of novel polymorphisms in the interleukin-6 (IL-6) gene on IL-6 transcription and plasma IL-6 levels, and an association with systemic-onset juvenile chronic arthritis. J Clin Invest 1998;102:1369–1376.

107 Burger MC, de Wet H, Collins M: Interleukin and growth factor gene variants and risk of carpal tunnel syndrome. Gene 2015;564:67–72.

108 Tworek D, Kuna P: The role of interleukin 10 in allergic inflammation (in Polish). Pol Merkur Lekarski 2005;18:125–128.

109 Lin WP, Lin JH, Chen XW, Wu CY, Zhang LQ, Huang ZD, et al: Interleukin-10 promoter polymorphisms associated with susceptibility to lumbar disc degeneration in a Chinese cohort. Genet Mol Res 2011;10:1719–1727.

110 Claudino M, Trombone APF, Cardoso CR, Ferreira SB, Martins W, Assis GF, et al: The broad effects of the functional IL-10 promoter -592 polymorphism: modulation of IL-10, TIMP-3, and OPG expression and their association with periodontal disease outcome. J Leukoc Biol 2008;84:1565–1573.

111 Haralambieva IH, Ovsyannikova IG, Dhiman N, Kennedy RB, O'Byrne M, Pankratz VS, et al: Common SNPs/haplotypes in *IL18R1* and *IL18* genes are associated with variations in humoral immunity to smallpox vaccination in Caucasians and African Americans. J Infect Dis 2011;204:433–441.

112 Omair A, Lie BA, Reikeras O, Brox JI: An association study of interleukin 18 receptor genes *(IL18R1* and *IL18RAP)* in lumbar disc degeneration. Open Orthop J 2012;6:164–171.

113 Hunt KA, Zhernakova A, Turner G, Heap GAR, Franke L, Bruinenberg M, et al: Newly identified genetic risk variants for celiac disease related to the immune response. Nat Genet 2008;40:395–402.

114 Omair A, Holden M, Lie BA, Reikeras O, Brox JI: Treatment outcome of chronic low back pain and radiographic lumbar disc degeneration are associated with inflammatory and matrix degrading gene variants: a prospective genetic association study. BMC Musculoskelet Disord 2013;14:105.

115 Pufe T, Petersen W, Tillmann B, Mentlein R: The angiogenic peptide vascular endothelial growth factor is expressed in foetal and ruptured tendons. Virchows Arch 2001;439:579–585.

116 Beye JA, Hart DA, Bray RC, McDougall JJ, Salo PT: Injury-induced changes in mRNA levels differ widely between anterior cruciate ligament and medial collateral ligament. Am J Sports Med 2008;36: 1337–1346.

117 Petersen W, Pufe T, Zantop T, Tillmann B, Tsokos M, Mentlein R: Expression of VEGFR-1 and VEGFR-2 in degenerative Achilles tendons. Clin Orthop Relat Res 2004;420:286–291.

118 Neufeld G, Cohen T, Gengrinovitch S, Poltorak Z: Vascular endothelial growth factor (VEGF) and its receptors. FASEB J 1999;13:9–22.
119 Petersen W, Varoga D, Zantop T, Hassenpflug J, Mentlein R, Pufe T: Cyclic strain influences vascular endothelial growth factor (VEGF) and the hypoxia inducible factor 1 alpha (HIF-1alpha) in tendon fibroblasts. J Orthop Res 2004;22:847–853.
120 Rahim M, Gibbon A, Hobbs H, van der Merwe W, Posthumus M, Collins M, et al: The association of genes involved in the angiogenesis-associated signaling pathway with risk of anterior cruciate ligament rupture. J Orthop Res 2014;32:1612–1618.
121 Shahbazi M, Fryer AA, Pravica V, Brogan IJ, Ramsay HM, Hutchinson IV, et al: Vascular endothelial growth factor gene polymorphisms are associated with acute renal allograft rejection. J Am Soc Nephrol 2002;13:260–264.
122 Lambrechts D, Storkebaum E, Morimoto M, Del-Favero J, Desmet F, Marklund SL, et al: VEGF is a modifier of amyotrophic lateral sclerosis in mice and humans and protects motoneurons against ischemic death. Nat Genet 2003;34:383–394.
123 Wang H, Keiser JA: Vascular endothelial growth factor upregulates the expression of matrix metalloproteinases in vascular smooth muscle cells : role of flt-1. Circ Res 1998;83:832–840.
124 Wang Y, Zheng Y, Zhang W, Yu H, Lou K, Zhang Y, et al: Polymorphisms of *KDR* gene are associated with coronary heart disease. J Am Coll Cardiol 2007;50:760–767.
125 Kim C-H: Early expression of myocardial HIF-1alpha in response to mechanical stresses: regulation by stretch-activated channels and the phosphatidylinositol 3-kinase signaling pathway. Circ Res 2002;90:25e–33e.
126 Lin W-P, Wang X-J, Wang C-R, Zhang L-Q, Li N, Wang F-S, et al: Polymorphism in the hypoxia-inducible factor 1alpha gene may confer susceptibility to LDD in Chinese cohort. PLoS One 2013; 8:e73158.
127 Tanimoto K, Yoshiga K, Eguchi H, Kaneyasu M, Ukon K, Kumazaki T, et al: Hypoxia-inducible factor-1alpha polymorphisms associated with enhanced transactivation capacity, implying clinical significance. Carcinogenesis 2003;24:1779–1783.
128 Buxton P, Edwards C, Archer CW, Francis-West P: Growth differentiation factor-5 (GDF-5) and skeletal development. J Bone Joint Surg Am 2001; 83-A(suppl):S23–S30.
129 Francis-West PH, Abdelfattah A, Chen P, Allen C, Parish J, Ladher R, et al: Mechanisms of GDF-5 action during skeletal development. Development 1999;126:1305–1315.
130 Mikic B, Schalet BJ, Clark RT, Gaschen V, Hunziker EB: GDF-5 deficiency in mice alters the ultrastructure, mechanical properties and composition of the Achilles tendon. J Orthop Res 2001;19:365–371.
131 Southam L, Rodriguez-Lopez J, Wilkins JM, Pombo-Suarez M, Snelling S, Gomez-Reino JJ, et al: An SNP in the 5′-UTR of *GDF5* is associated with osteoarthritis susceptibility in Europeans and with in vivo differences in allelic expression in articular cartilage. Hum Mol Genet 2007;16:2226–2232.
132 Posthumus M, Collins M, Cook J, Handley CJ, Ribbans WJ, Smith RKW, et al: Components of the transforming growth factor-beta family and the pathogenesis of human Achilles tendon pathology – a genetic association study. Rheumatology 2010;49:2090–2097.
133 Williams FMK, Popham M, Hart DJ, de Schepper E, Bierma-Zeinstra S, Hofman A, et al: *GDF5* single-nucleotide polymorphism rs143383 is associated with lumbar disc degeneration in Northern European women. Arthritis Rheum 2011;63:708–712.
134 Raleigh SM, Posthumus M, O'Cuinneagain D, van der Merwe W, Collins M: The *GDF5* gene and anterior cruciate ligament rupture. Int J Sports Med 2012;34:364–367.
135 Jurutka PW, Remus LS, Whitfield GK, Thompson PD, Hsieh JC, Zitzer H, et al: The polymorphic N terminus in human vitamin D receptor isoforms influences transcriptional activity by modulating interaction with transcription factor IIB. Mol Endocrinol 2000;14:401–420.
136 Videman T, Leppävuori J, Kaprio J, Battié MC, Gibbons LE, Peltonen L, et al: Intragenic polymorphisms of the vitamin D receptor gene associated with intervertebral disc degeneration. Spine 1998; 23:2477–2485.
137 Videman T, Gibbons LE, Battié MC, Maravilla K, Vanninen E, Leppävuori J, et al: The relative roles of intragenic polymorphisms of the vitamin D receptor gene in lumbar spine degeneration and bone density. Spine 2001;26:E7–E12.
138 Cheung KMC, Orth F, Chan D, Karppinen J, Chen Y, Jim JJT, et al: Association of the *Taq* I allele in vitamin D receptor with degenerative disc disease and disc bulge in a Chinese population 2006;31: 1143–1148.
139 Kawaguchi Y, Kanamori M, Ishihara H, Ohmori K, Matsui H, Kimura T: The association of lumbar disc disease with vitamin-D receptor gene polymorphism. J Bone Joint Surg Am 2002;84-A:2022–2028.
140 Jurutka PW, Whit GK, Hsieh J, Thompson PD, Haussler CA, Haussler MR: Molecular nature of the vitamin D receptor and its role in regulation of gene expression. Rev Endocr Metab Disord 2001;2:203–216.

141 Lian Ø, Scott A, Engebretsen L, Bahr R, Duronio V, Khan K: Excessive apoptosis in patellar tendinopathy in athletes. Am J Sports Med 2007;35:605–611.
142 Yuan J, Murrell GA, Wei A-Q, Wang M-X: Apoptosis in rotator cuff tendinopathy. J Orthop Res 2002;6:2–9.
143 Scott A, Khan KM, Heer J, Cook JL, Lian O, Duronio V: High strain mechanical loading rapidly induces tendon apoptosis: an ex vivo rat tibialis anterior model. Br J Sports Med 2005;39:e25.
144 Nell E-M, van der Merwe L, Cook J, Handley CJ, Collins M, September AV: The apoptosis pathway and the genetic predisposition to Achilles tendinopathy. J Orthop Res 2012;30:1719–1724.
145 Sun T, Gao Y, Tan W, Ma S, Shi Y, Yao J, et al: A six-nucleotide insertion-deletion polymorphism in the *CASP8* promoter is associated with susceptibility to multiple cancers. Nat Genet 2007;39:605–613.
146 Rannou F, Lee T-S, Zhou R-H, Chin J, Lotz JC, Mayoux-Benhamou M-A, et al: Intervertebral disc degeneration: the role of the mitochondrial pathway in annulus fibrosus cell apoptosis induced by overload. Am J Pathol 2004;164:915–924.
147 Park J-B, Lee J-K, Park S-J, Kim K-W, Riew KD: Mitochondrial involvement in Fas-mediated apoptosis of human lumbar disc cells. J Bone Joint Surg Am 2005;87:1338–1342.
148 Tschoeke SK, Hellmuth M, Hostmann A, Robinson Y, Ertel W, Oberholzer A, et al: Apoptosis of human intervertebral discs after trauma compares to degenerated discs involving both receptor-mediated and mitochondrial-dependent pathways. J Orthop Res 2008;26:999–1006.
149 Sun Z-M, Ling M, Huo Y, Chang Y, Li Y, Qin H, et al: Caspase 9 gene polymorphism and susceptibility to lumbar disc disease in the Han population in northern China. Connect Tissue Res 2011;52:198–202.
150 Andreoli V, Trecroci F, La Russa A, Valentino P, Condino F, Latorre V, et al: CASP-9: a susceptibility locus for multiple sclerosis in Italy. J Neuroimmunol 2009;210:100–103.
151 Park J, Kim K, Han C, Chang H: Expression of Fas receptor on disc cells in herniated lumbar disc tissue. Spine 2001;26:142–146.
152 Park J, Chang H, Kim K: Expression of Fas ligand and apoptosis of disc cells in herniated lumbar disc tissue. Spine 2001;26:618–621.
153 Zhu G-B, Jiang X-R, Xia C-L, Sun Y-J, Zeng Q-S, Wu X-M, et al: Association of *FAS* and *FAS* ligand polymorphisms with the susceptibility and severity of lumbar disc degeneration in Chinese Han population. Biomarkers 2011;16:485–490.
154 Wu J, Metz C, Xu X, Abe R, Gibson AW, Edberg JC, et al: A novel polymorphic CAAT/enhancer-binding protein beta element in the FasL gene promoter alters Fas ligand expression: a candidate background gene in African American systemic lupus erythematosus patients. J Immunol 2003;170:132–138.
155 Zhou Z-X, Mi Y-Y, Ma HZ, Zou J-G, Zhang L-F: *FAS* –1377 G/A (rs2234767) polymorphism and cancer susceptibility: a meta-analysis of 17,858 cases and 24,311 controls. PLoS One 2013;8:e73700.
156 Shang X-P, Sun X-C, Wang Y-X, Ju B-B: Association of BCL-2 polymorphism with the presence and severity of lumbar disc degeneration in the Chinese Han population. Clin Lab 2012;58:261–266.
157 Raleigh SM, Collins M: Gene variants that predispose to Achilles tendon injuries: an update on recent advances; in Cretnik A (ed): Achilles Tendon. Rejika, InTech, 2012.
158 September A, Posthumus M, Collins M: Application of genomics in the prevention, treatment and management of Achilles tendinopathy and anterior cruciate ligament ruptures. Recent Pat DNA Gene Seq 2012;6:216–223.
159 Ribbans WJ, Collins M: Pathology of the tendo Achillis: do our genes contribute? Bone Joint J 2013;95-B:305–313.
160 Ohi T, Uehara Y, Takatsu M, Watanabe M, Ono T: Hypermethylation of CpGs in the promoter of the *COL1A1* gene in the aged periodontal ligament. J Dent Res 2006;85:245–250.
161 Brown WM: Exercise-associated DNA methylation change in skeletal muscle and the importance of imprinted genes: a bioinformatics meta-analysis. Br J Sports Med 2015;49:1567–1578.

Assoc. Prof. Alison September
Division of Exercise Science and Sports Medicine
University of Cape Town
PO Box 115, Newlands 7725 (South Africa)
E-Mail alison.september@uct.ac.za

Posthumus M, Collins M (eds): Genetics and Sports, ed 2, revised, extended.
Med Sport Sci. Basel, Karger, 2016, vol 61, pp 92–104 (DOI: 10.1159/000445242)

Genetics of Musculoskeletal Exercise-Related Phenotypes

Malcolm Collins[a] · Kevin O'Connell[b] · Michael Posthumus[a]

[a]Division of Exercise Science and Sports Medicine, Department of Human Biology, Faculty of Health Sciences, University of Cape Town, Cape Town, and [b]Department of Genetics, Stellenbosch University, Stellenbosch, South Africa

Abstract
Genetic factors have been shown to associate with various exercise-related phenotypes, including exercise performance, adaptation to training and sports injuries. The genes implicated in the pathogenesis of musculoskeletal soft-tissue injuries all code for either structural components or regulatory components of the extracellular matrix. It has been hypothesized that these genetic associations with injuries are due to genetically regulated changes in mechanical properties of musculoskeletal soft tissue. Thus, the objective of this review is to highlight the research which has advanced our understanding of how genetic variation within these structural genes affects the properties of our connective tissue. The genetics of various exercise-related phenotypes, such as range of motion, endurance performance and exercise-associated muscle cramps, are reviewed. Lastly, a model is presented where genetic variations within a collagen-encoding gene result in a continuum of phenotype ranging from a normal tissue to a seriously deleterious or lethal disorder.

As highlighted in previous reviews of this book, genetic factors have been shown to associate with various exercise-related phenotypes, including exercise performance, adaptation to training and sports injuries. Rahim et al. [this vol., pp. 68–91] thoroughly reviewed the genetic loci which have been associated with musculoskeletal soft-tissue injuries. The genes implicated in the pathogenesis of musculoskeletal soft-tissue injuries all code for either structural components, or regulatory components of the extracellular matrix. Further, certain of the implicated genetic loci, mainly within the collagen-encoding genes, have also been associated with other exercise-related phenotypes, such as exercise performance, range of motion (ROM) and musculotendinous stiffness. Although several genetic loci, such as *COL3A1*, *COL6A1* and *COL12A1*, have been investigated, variants within the *COL5A1* gene have been the most extensively investigated. The objective of this review is to highlight the research which has advanced our understanding of how genetic variation within these structural genes affects the properties of our connective tissue. Therefore, we will outline the basic collagen fibril structure and its relation to mechanical properties of the musculoskeletal tissue, followed by a review of previous genetic associations with the collagen-encoding genes. Lastly, as an explanation of the reviewed associations, it is discussed, by highlighting the advances in understanding the mechanism underlying the *COL5A1* gene regulation, how structural genes may have such vast effects on musculoskeletal soft tissue.

Collagen Fibril Organization and the Mechanical Properties of Musculoskeletal Soft Tissue

Collagen fibrils, which are arranged in a highly organized hierarchical structure [1], form the basic structural component of tendons, ligaments and connective-tissue structures in skeletal muscle [2]. Specifically, these collagen fibrils account for approximately 85% of the dry weight of the extracellular matrix of tendons, while other non-collagen components include, amongst others, 1–5% proteoglycans and 2% elastin [3]. The collagen fibril consists predominately of type I collagen (95–99%) and type III collagen (1–5%) [1, 3]. Other minor collagens within the tendon include, amongst others, types IV, V, VI, XII and XIV (table 1) [1, 3]. Expression of these collagens may be specific to portions of the tendon (table 1), including the tendon midsubstance, myotendinous junction and osteotendinous junction (enthuses).

Although there are important functional and macromolecular composition differences, ligaments and tendons have a very similar hierarchical structure [1]. Unlike tendons, only 70–80% of the dry mass of ligaments is made up of collagen fibres, while the remainder of the ligament extracellular matrix is composed of glycoproteins, proteoglycans and glycosaminoglycans [1]. The collagen composition of ligaments is approximately 90% type I collagen, 10% type III collagen and as much as 12% type V collagen [1] (table 1).

Rare mutations in many collagen-encoding genes result in serious musculoskeletal disorders [4–12], and symptoms of these disorders include, but are not limited to, bone fragility, joint hypermobility, skin hyperextensibility and/or abnormal wound healing [5, 9]. The nature of these symptoms highlights the importance of collagen in maintaining the normal architecture and mechanical properties of musculoskeletal soft tissues. Common polymorphisms within collagen-encoding genes have also been associated with several multifactorial disorders [13–30], including susceptibility to

Table 1. A summary of 'well-characterized' collagens identified within tendons, ligaments and the connective-tissue structures of skeletal muscle

Collagen types	Tendon			Ligament	Skeletal muscle		
	midsub.	MTJ	OTJ		endo.	peri.	epi.
Fibrillar							
I	√	√	√	√	√	√	√
II			√				
III	√		√	√	√	√	√
V	√			√	√	√	
XI				√			
Non-fibrillar							
IV		√			√		
VI		√	√	√	√		
XII	√			√		√	
XIV	√			√		√	

The localization of collagen types is indicated by 'ticks'. The tendon is shown as 3 separate portions: the midsubstance (midsub.), the myotendinous junction (MTJ) and the osteotendinous junction (OTJ). The 3 separate connective-tissue subdivisions of skeletal muscle are shown: the endomysium (endo.), the perimysium (peri.) and the epimysium (epi.).

musculoskeletal injuries [13–24]. For example, rare mutations within *COL5A1* cause Ehlers-Danlos syndrome [7–9], while common polymorphisms within this gene have been associated with chronic Achilles tendinopathy [14, 15]. Investigation of haplo-insufficient *Col5a1* +/– mice, in comparison to wild-type mice, showed that reduced levels of type V collagen resulted in greater numbers of both larger and smaller collagen fibrils in the dermis [31]. These haplo-insufficient mice also showed a reduced dermis fibril density when compared to the wild-type mice [31]. Similar phenotypes have been observed within the murine tendon when type V and type XI collagen interactions were investigated [32]. It is therefore not surprising that the collagen genes have been identified as ideal candidates for the association with susceptibility to injury and other exercise-associated phenotypes that involve the musculoskeletal system [33].

Structural and architectural changes to the collagen fibrils are known to alter the biomechanical properties of their containing tissues [34, 35]. The measures used to describe the biomechanical properties of these tissues include ultimate tensile strength, Young's modulus and creep [36]. The ultimate tensile strength refers to the maximum stress that the tissue can take before failing [36]. Young's modulus, a measure of stiffness, is defined as the ratio of stress (force per unit area) over strain (deformation relative to initial length) along a particular axis [36]. Finally, creep is the long-term deformation of a tissue due to mechanical stress [36].

An increased mean collagen fibril diameter results in greater ultimate tensile strength [35, 37]. Furthermore, the larger diameter also increases the surface area per collagen fibril resulting in an increased number of covalent cross-links between the adjacent fibrils, thereby proposed to inhibit creep [35, 37]. A smaller mean collagen fibril diameter results in increased fibril density within the fibre and a reduced surface area per collagen fibril [35, 37]. Although the surface area per fibril is reduced, the increased density of smaller fibrils results in a larger overall surface area per unit mass [35, 37]. This overall increase in surface area is proposed to lead to increased creep, and possibly stiffness of the tissue, due to increased electrostatic interactions between the fibrils and ground substance [35].

Collagen Structural Genes and Exercise-Related Phenotypes

Type V collagen is known to interact with types I and III and plays a role in the regulation of collagen fibrillogenesis [38, 39]. As reviewed in Rahim et al. [this vol., pp. 68–91], variants within the 3′-untranslated region (UTR) of the *COL5A1* gene, which encodes the α_1-chain of type V collagen, has been associated with modulating the risk of chronic Achilles tendinopathy [14, 15], tennis elbow [40] and anterior cruciate ligament ruptures in females [23]. Specifically, the CC genotype of rs12722 (T/C) was associated with a reduced risk of injury, while the rs71746744 AGGG/AGGG, rs169399 deletion/deletion (or –/–) and rs1134170 TT genotypes are associated with an increased risk of chronic Achilles tendinopathy [14, 15, 23, 40].

Since the 3′-UTR of eukaryotic genes contains elements which are emerging as important posttranscriptional regulators [41, 42], it has been suggested that the *COL5A1* 3′-UTR plays a role in the regulation of the *COL5A1* mRNA stability and by implication type V collagen production. In support of this, two major functional allelic forms of the *COL5A1* 3′-UTR, namely the C and T functional forms, have been identified and show significant differences in mRNA stability, with the T functional form having increased stability [43]. The 4 *COL5A1* 3′-UTR alleles, namely the rs12722 T, rs71746744 AGGG, rs169399 deletion (or –) and rs1134170 T alleles, associated with an increased risk of chronic Achilles tendinopathy have been identified in the T functional form, while the opposite alleles are all present in the C functional form [43]. Based on these findings, Collins and Posthumus [44] proposed that the relative content of type V collagen in tendons, ligaments and other tissues alters the fibril diameters and packing density within these tissues. This will alter their mechanical properties and therefore their susceptibility to injury and other exercise-related phenotypes. The *COL5A1* 3′-UTR variants have therefore also been investigated for associations with other exercise-related phenotypes, including ROM measurements [45–47], endurance performance [45, 48, 49] and exercise-associated muscle cramps [50]. The following subsections will specifically review the role of *COL5A1*, as well as other collagen genes, within these specific phenotypes. Where other collagen genes have also been associated, these will also be included with a brief description of their role, where appropriate.

Range of Motion

Altered musculotendinous flexibility, which is defined as 'the ability to move a joint through its complete range of motion', has been cited as an intrinsic risk factor for many injuries [51–53], including tendon and ligament injuries. Although flexibility can be improved through regular stretching, it is at least in part also determined genetically [54]. Since mutations within *COL5A1* cause classic Ehlers-Danlos syndrome, a condition that is characterized by joint hypermobility and other connective tissue-related symptoms [9], the association of variants within the *COL5A1* 3′-UTR with ROM has been investigated [45–47, 49, 55–57].

The *COL5A1* rs12722 (T/C) variant was associated with lower limb ROM measurements in a mixed injured/uninjured cohort [47]. In addition, a follow-up study in an independent, larger cohort of apparently healthy physically active Caucasian South Africans reported that the *COL5A1* rs12722 CC genotype protected participants against an age-related decline in sit-and-reach ROM [46]. When divided into two age groups, the *COL5A1* rs12722 CC genotype was significantly associated with increased sit-and-reach ROM in the older (>35 years), but not younger (<35 years) age group [46]. Similarly, a recent study reported that the rs12722 genotype was not associated with sit-and-reach ROM measurements in a young (25.2 ± 4.0 years) Brazilian cohort [55]. Although sit-and-reach ROM was not measured in the study, the *COL5A1* rs12722 CC genotype was reported to associate with increased straight leg raise ROM in a young (23.4 ± 1.2 years) Asian population, consisting of Korean and Japanese college students [56]. These authors also reported a tendency for the CC genotype to associate with increased whole-body joint laxity. The *COL5A1* rs12722 CT genotype has also been associated with high joint mobility, the occurrence of genu recurvatum and a higher incidence of injuries in a cohort of Italian high-level international rhythmic gymnasts, further implicating this variant in the modulation of joint mobility [57].

Bell et al. [58] investigated whether an association exists between the *COL5A1* rs12722 variant and anterior knee laxity, genu recurvatum and generalized joint laxity. The CT genotype was associated with increased genu recurvatum and general joint laxity. Further, the *COL1A1* and *COL12A1* genes were also investigated for their association with these laxity phenotypes.

Type I collagen, the major fibrillar collagen in ligaments and tendons, is a heterotrimer consisting of 2 α_1(I)- and 1 α_2(I)-chains, which are encoded for by the *COL1A1* and *COL1A2* genes, respectively [59]. A functional Sp1 binding site polymorphism within the *COL1A1* gene has also been shown to be associated with several multifactorial disorders [60, 61] and a joint hypermobility disorder, Ehlers-Danlos syndrome [62]. The TT of the genotype of the Sp1 binding site polymorphism was found to be associated with greater knee hyperextension measurements [58]. However, Khoschnau et al. [63] found no association between this genotype and generalized joint laxity.

Type XII is involved in a biological process similar to type V collagen: both collagens are believed to regulate microfibril diameter (fibrillogenesis) [38, 39, 64]. Type XII collagen is a homotrimer consisting of 3 α_1(XII)-chains and is encoded for by a single gene, *COL12A1* [65]. Bell et al. [58] found an association between the *COL12A1* rs970547 variant and anterior knee laxity and generalized joint laxity. The AA genotype of rs970547 was found to be associated with greater anterior knee laxity in females only, whereas the CC genotype was associated with a greater general joint laxity in males [58]. There were however no significant associations between the *COL12A1* rs970547 variant and measurements of ROM in a South African cohort [66].

Although rare mutations within other collagen-encoding genes, such as *COL3A1* and *COL6A1*, also cause connective-tissue hypermobility disorders and phenotypes, none of the common variants within these genes have to date been reported to associate with ROM measurements [66].

Endurance Performance

Reduced ROM, specifically sit-and-reach ROM, has also been associated with improved endurance running economy and performance [67, 68]. This relationship is proposed to result from reduced aerobic demand due to increased elastic return and may result in improved performance [68]. Given that the *COL5A1* rs12722 variant is associated with sit-and-reach ROM [46, 47], it is not surprising that this variant was also associated with endurance running performance. The TT genotype of this variant was associated with faster run times in South African Ironman triathletes while triathletes with the CC genotype were overrepresented in the slowest tertile for the run stage of this triathlon [48]. The TT genotype was also associated with faster performance in a 56-km ultramarathon road race [45]. Furthermore, when these endurance runners were divided into 4 quadrants (fast and inflexible, fast and flexible, slow and inflexible, slow and flexible) based on both their race finishing times and sit-and-reach ROM measures, runners with the *COL5A1* rs12722 T allele were overrepresented in the fast and inflexible quadrant. Although it is believed that these findings may be mediated by changes to the energetic cost of running, there were no significant associations between the *COL5A1* rs12722 genotype and running economy [55].

A second *COL5A1* variant located within the functional *COL5A1* 3′-UTR, rs71746744 (–/AGGG), was later shown to be associated with both endurance running performance and sit-and-reach ROM measures in 56-km ultramarathon runners [49]. Specifically, the rs71746744 AGGG/AGGG genotype was associated with increased performance and reduced sit-and-reach ROM, similar to the rs12722 TT genotype. Interestingly the rs71746744 AGGG and rs12722 T alleles contribute to the functional T form as described above.

Similarly, the *COL3A1* rs1800255 (G/A), *COL12A1* rs970547 (A/G) and *COL5A1* rs12722 variants interact to modulate endurance running performance in Ironman triathlons [69]. Interestingly, although not independently associated with running performance, the *COL6A1* rs35796750 (T/C) TT genotype was associated with faster cycling times during the Ironman triathlon [70].

Type III collagen, which is also implicated in fibrillogenesis and is a major fibrillar collagen which forms heterotypic fibrils together with type I collagen, is a homotrimer consisting of 3 α_1(III)-chains encoded by *COL3A1* [71–74]. The functional non-synonymous *COL3A1* rs1800255 variant is also associated with a mitral valve and pelvic organ prolapse [28–30]. The alanine-to-threonine change at position 698 of the α_1(III)-chain, as a result of *COL3A1* rs1800255, is proposed to affect the tensile strength of type III collagen fibres [28].

Although the function of type VI collagen remains largely unknown, it is believed to play a role at the basement membrane [39, 75] and is known to interact with type V collagen in the fibril [76]. Mutations within the gene which encodes the α_1-chain of type VI collagen (*COL6A1*) have been shown to cause muscle diseases such as Bethlem myopathy [75, 77] and Ullrich congenital muscular dystrophy [75, 77, 78]. Bonaldo et al. [79] established a model for myopathy in mice by knocking out the *Col6a1* gene. The average weekly distance run by these myopathic *Col6a1* –/– mice was consistently less than that of the wild-type mice [79]. In addition, these *Col6a1* null mice showed abnormal fibril diameter and density [80], similar to that observed in *Col5a1* haploinsufficient mice [31]. Sequence variations, a variant within intron 32 (rs35796750 T/C), in the human *COL6A1* gene, have to date been associated with a number of multifactorial conditions, such as ossification of the posterior longitudinal ligament [25, 81], ossification of the ligamentum flavum [25] and diffuse idiopathic skeletal hyperostosis [27]. This variant occurs near the branch site of intron 32, where the cytosine-to-thymine transition is proposed to cause aberrant splicing of the *COL6A1* mRNA [81].

Exercise-Associated Muscle Cramps

Exercise-associated muscle cramping (EAMC) is a common medical condition among participants of endurance events such as ultramarathons and triathlons [82], and one hypothesis implicates (i) an increased exercise intensity or duration [83–85], (ii) the development of muscle fatigue [84], (iii) contraction of the muscle in a shortened position [84] and (iv) possible tissue damage [84] as the primary factors in the development. There is also evidence of a familial predisposition to EAMC [83]. Therefore, the *COL5A1* rs12722 variant was investigated in the aetiology of EAMC. Similarly to other musculoskeletal soft-tissue injuries, the CC genotype was significantly overrepresented in South African Ironman triathlon participants with no self-reported history of previous (lifelong) EAMC when compared to participants who reported a history of EAMC within 12 months prior to the event [50].

Conclusion and Future Work

As mentioned above, murine and in vitro models have shown the effects that collagen gene mutations may have on musculoskeletal soft tissue, including aberrant changes to collagen fibril size or diameter and fibril density. Therefore, it may be proposed that less severe changes to the musculoskeletal soft-tissue architecture may result in other phenotypic effects. For example, changes to the fibril architecture and organization as a result of functional variants within the *COL5A1* gene may explain associations identified between this gene and musculoskeletal soft-tissue injuries and other exercise-related phenotypes [44]. Specifically, smaller fibrils with greater density may result in reduced ROM, improved endurance running performance and reduced risk of musculoskeletal soft-tissue injuries [44]. In support of this, one recent in vivo analysis of *COL5A1* rs12722 (T/C) in Japanese males showed that individuals with the CC genotype had greater mean maximal tendon elongation (TT + TC 21.1 ± 5.4 mm, CC 24.5 ± 5.4 mm) and strain (TT + TC 6.51 ± 1.58%, CC 7.61 ± 1.62%), and lower stiffness (TT + TC 78.2 ± 18.5 N/mm, CC 66.2 ± 19.3 N/mm), for their knee extensors when compared to individuals with the TT or TC genotypes [86]. No significant differences were recorded for the plantar flexors [86]. This study highlights that the level of effect of gene variants may differ between tissue structures, possibly due to differences in environmental stimuli, and adds further evidence to the proposed genetic continuum model updated in figure 1 [44, 86].

A second in vivo investigation in male and female Caucasian participants identified no associations between the *COL5A1* rs12722 genotypes and patellar tendon or plantar flexor architecture and mechanical properties, including maximal elongation, strain and stiffness [87]. Interestingly, although not significant, the median maximal stiffness showed a tendency to be higher in the TT genotype group (855.7 N/mm), followed by the TC genotype (707.6 N/mm), while the CC genotype group (555.3 N/mm) reported the lowest maximal stiffness. Differences between the combined TT + TC genotypes compared to the CC genotype group were however not reported. A number of methodological disparities between these studies may explain the differences observed. These include ethnic and gender differences between study populations, as well as the techniques and methods used to determine the dimensional and functional properties of the tendons in vivo [86, 87]. Further studies are therefore required to determine the exact effects that *COL5A1* rs12722 may have on tendon architecture and mechanical properties, as well as further investigate further variants, which may interact with *COL5A1*. For example, genes encoding structural and functionally related type XI *(COL11A1 and COL11A2)* and type V *(COL5A1)* collagens have been shown to interact with one another to collectively modulate the risk for Achilles tendinopathy. This haplotype may further modulate the mechanical properties of musculoskeletal tissue.

Despite the differences in these in vivo tendon analyses, it is known that altered type V collagen expression affects collagen fibril architecture and organization [31,

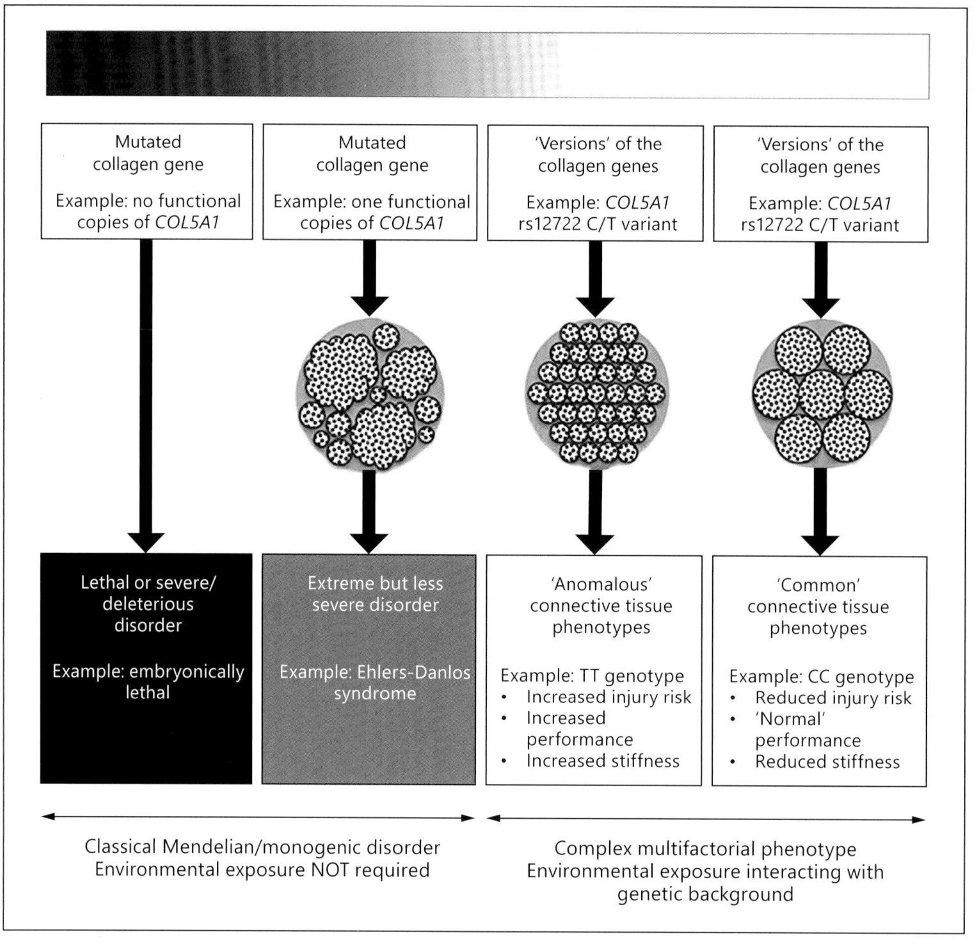

Fig. 1. A proposed general genetic continuum for collagen genes, adapted from Collins and Posthumus [44]. The black shading represents the lethal or severe phenotypes due to mutations in the collagen genes. At this end of the continuum, a single mutation results in the disorder. Murine and in vitro models have shown the effects that collagen gene mutations may have on musculoskeletal soft tissue, including aberrant changes to collagen fibril size or diameter and fibril density [32]. The white shading represents the most beneficial 'versions' of the collagen genes. At this end of the continuum, variants within collagen genes collectively contribute to the aetiology of the phenotype. Smaller fibrils with greater density, as a result of variants such as rs12722, may result in reduced ROM, improved endurance running performance and increased risk of musculoskeletal soft-tissue injuries [44, 86].

32]. Like type V collagen, changes to types XII, III and VI collagen levels also result in changes to the fibril architecture and organization which may affect the biomechanical properties of tissues containing these proteins [79, 80, 88, 89]. Therefore, it may be proposed that common functional variants, like *COL5A1* rs12722, within the genes that encode these proteins may be associated with musculoskeletal soft-tissue injuries and other exercise-related phenotypes.

References

1 Hoffmann A, Gross G: Tendon and ligament engineering in the adult organism: mesenchymal stem cells and gene-therapeutic approaches. Int Orthop 2007;31:791–797.

2 O'Connell K, Saunders CJ, Collins M: Collagen gene sequence variants in exercise-related traits. Cent Eur J Sport Sci Med 2013;1:3–17.

3 Jozsa L, Kannus P: Human Tendons: Anatomy, Physiology and Pathology. Champaign, Human Kinetics, 1997.

4 Pace JM, Corrado M, Missero C, Byers PH: Identification, characterization and expression analysis of a new fibrillar collagen gene, *COL27A1*. Matrix Biol J Int Soc Matrix Biol 2003;22:3–14.

5 Pollitt R, McMahon R, Nunn J, Bamford R, Afifi A, Bishop N, et al: Mutation analysis of *COL1A1* and *COL1A2* in patients diagnosed with osteogenesis imperfecta type I–IV. Hum Mutat 2006;27:716.

6 Warman ML, Abbott M, Apte SS, Hefferon T, McIntosh I, Cohn DH, et al: A type X collagen mutation causes Schmid metaphyseal chondrodysplasia. Nat Genet 1993;5:79–82.

7 Malfait F, De Paepe A: Molecular genetics in classic Ehlers-Danlos syndrome. Am J Med Genet C Semin Med Genet 2005;139C:17–23.

8 Malfait F, Coucke P, Symoens S, Loeys B, Nuytinck L, De Paepe A: The molecular basis of classic Ehlers-Danlos syndrome: a comprehensive study of biochemical and molecular findings in 48 unrelated patients. Hum Mutat 2005;25:28–37.

9 Malfait F, Wenstrup RJ, De Paepe A: Clinical and genetic aspects of Ehlers-Danlos syndrome, classic type. Genet Med Off J Am Coll Med Genet 2010;12: 597–605.

10 Byers PH, Duvic M, Atkinson M, Robinow M, Smith LT, Krane SM, et al: Ehlers-Danlos syndrome type VIIA and VIIB result from splice-junction mutations or genomic deletions that involve exon 6 in the *COL1A1* and *COL1A2* genes of type I collagen. Am J Med Genet 1997;72:94–105.

11 Smith LT, Schwarze U, Goldstein J, Byers PH: Mutations in the *COL3A1* gene result in the Ehlers-Danlos syndrome type IV and alterations in the size and distribution of the major collagen fibrils of the dermis. J Invest Dermatol 1997;108:241–247.

12 Zou Y, Zwolanek D, Izu Y, Gandhy S, Schreiber G, Brockmann K, et al: Recessive and dominant mutations in *COL12A1* cause a novel EDS/myopathy overlap syndrome in humans and mice. Hum Mol Genet 2014;23:2339–2352.

13 Abrahams Y, Laguette M-J, Prince S, Collins M: Polymorphisms within the *COL5A1* 3′-UTR that alters mRNA structure and the *MIR608* gene are associated with Achilles tendinopathy. Ann Hum Genet 2013;77:204–214.

14 Mokone GG, Schwellnus MP, Noakes TD, Collins M: The *COL5A1* gene and Achilles tendon pathology. Scand J Med Sci Sports 2006;16:19–26.

15 September AV, Cook J, Handley CJ, van der Merwe L, Schwellnus MP, Collins M: Variants within the *COL5A1* gene are associated with Achilles tendinopathy in two populations. Br J Sports Med 2009;43: 357–365.

16 Hay M, Patricios J, Collins R, Branfield A, Cook J, Handley CJ, et al: Association of type XI collagen genes with chronic Achilles tendinopathy in independent populations from South Africa and Australia. Br J Sports Med 2013;47:569–574.

17 Posthumus M, September AV, Schwellnus MP, Collins M: Investigation of the Sp1-binding site polymorphism within the *COL1A1* gene in participants with Achilles tendon injuries and controls. J Sci Med Sport Sports Med Aust 2009;12:184–189.

18 Saunders CJ, van der Merwe L, Posthumus M, Cook J, Handley CJ, Collins M, et al: Investigation of variants within the *COL27A1* and *TNC* genes and Achilles tendinopathy in two populations. J Orthop Res Off Publ Orthop Res Soc 2013;31:632–637.

19 September AV, Posthumus M, van der Merwe L, Schwellnus M, Noakes TD, Collins M: The *COL12A1* and *COL14A1* genes and Achilles tendon injuries. Int J Sports Med 2008;29:257–263.

20 Ficek K, Cieszczyk P, Kaczmarczyk M, Maciejewska-Karłowska A, Sawczuk M, Cholewinski J, et al: Gene variants within the *COL1A1* gene are associated with reduced anterior cruciate ligament injury in professional soccer players. J Sci Med Sport Sports Med Aust 2013;16:396–400.

21 O'Connell K, Knight H, Ficek K, Leonska-Duniec A, Maciejewska-Karlowska A, Sawczuk M, et al: Interactions between collagen gene variants and risk of anterior cruciate ligament rupture. Eur J Sport Sci 2015;15:341–350.

22 Posthumus M, September AV, Keegan M, O'Cuinneagain D, Van der Merwe W, Schwellnus MP, et al: Genetic risk factors for anterior cruciate ligament ruptures: *COL1A1* gene variant. Br J Sports Med 2009;43:352–356.

23 Posthumus M, September AV, O'Cuinneagain D, van der Merwe W, Schwellnus MP, Collins M: The *COL5A1* gene is associated with increased risk of anterior cruciate ligament ruptures in female participants. Am J Sports Med 2009;37:2234–2240.

24 Posthumus M, September AV, O'Cuinneagain D, van der Merwe W, Schwellnus MP, Collins M: The association between the *COL12A1* gene and anterior cruciate ligament ruptures. Br J Sports Med 2010;44: 1160–1165.

25 Kong Q, Ma X, Li F, Guo Z, Qi Q, Li W, et al: *COL6A1* polymorphisms associated with ossification of the ligamentum flavum and ossification of the posterior longitudinal ligament. Spine 2007;32:2834–2838.
26 Ikegawa S: Genetics of ossification of the posterior longitudinal ligament of the spine: a mini review. J Bone Metab 2014;21:127–132.
27 Tsukahara S, Miyazawa N, Akagawa H, Forejtova S, Pavelka K, Tanaka T, et al: *COL6A1,* the candidate gene for ossification of the posterior longitudinal ligament, is associated with diffuse idiopathic skeletal hyperostosis in Japanese. Spine 2005;30:2321–2324.
28 Kluivers KB, Dijkstra JR, Hendriks JCM, Lince SL, Vierhout ME, van Kempen LCL: *COL3A1* 2209G>A is a predictor of pelvic organ prolapse. Int Urogynecol J Pelvic Floor Dysfunct 2009;20:1113–1118.
29 Chou H-T, Hung J-S, Chen Y-T, Wu J-Y, Tsai F-J: Association between *COL3A1* collagen gene exon 31 polymorphism and risk of floppy mitral valve/mitral valve prolapse. Int J Cardiol 2004;95:299–305.
30 Chen H-Y, Chung Y-W, Lin W-Y, Wang J-C, Tsai F-J, Tsai C-H: Collagen type 3 alpha 1 polymorphism and risk of pelvic organ prolapse. Int J Gynaecol Obstet Off Organ Int Fed Gynaecol Obstet 2008;103: 55–58.
31 Wenstrup RJ, Florer JB, Davidson JM, Phillips CL, Pfeiffer BJ, Menezes DW, et al: Murine model of the Ehlers-Danlos syndrome. col5a1 haploinsufficiency disrupts collagen fibril assembly at multiple stages. J Biol Chem 2006;281:12888–12895.
32 Wenstrup RJ, Smith SM, Florer JB, Zhang G, Beason DP, Seegmiller RE, et al: Regulation of collagen fibril nucleation and initial fibril assembly involves coordinate interactions with collagens V and XI in developing tendon. J Biol Chem 2011;286:20455–20465.
33 Ribbans WJ, Collins M: Pathology of the tendo Achillis: do our genes contribute? Bone Jt J 2013;95-B:305–313.
34 Kongsgaard M, Qvortrup K, Larsen J, Aagaard P, Doessing S, Hansen P, et al: Fibril morphology and tendon mechanical properties in patellar tendinopathy: effects of heavy slow resistance training. Am J Sports Med 2010;38:749–756.
35 Ottani V, Raspanti M, Ruggeri A: Collagen structure and functional implications. Micron 2001;32:251–260.
36 Maganaris CN, Narici MV, Maffulli N: Biomechanics of the Achilles tendon. Disabil Rehabil 2008;30: 1542–1547.
37 Parry DA, Barnes GRG, Craig AS: A comparison of the size distribution of collagen fibrils in connective tissues as a function of age and a possible relation between fibril size distribution and mechanical properties. Proc R Soc Lond B Biol Sci 1978;203:305–321.
38 Birk DE: Type V collagen: heterotypic type I/V collagen interactions in the regulation of fibril assembly. Micron 2001;32:223–237.
39 Van der Rest M, Garrone R: Collagen family of proteins. FASEB J 1991;5:2814–2823.
40 Altinisik J, Meric G, Erduran M, Ates O, Ulusal AE, Akseki D: The *Bst*UI and *Dpn*II variants of the *COL5A1* gene are associated with tennis elbow. Am J Sports Med 2015;43:1784–1789.
41 Mazumder B, Seshadri V, Fox PL: Translational control by the 3′-UTR: the ends specify the means. Trends Biochem Sci 2003;28:91–98.
42 Xie X, Lu J, Kulbokas EJ, Golub TR, Mootha V, Lindblad-Toh K, et al: Systematic discovery of regulatory motifs in human promoters and 3′-UTRs by comparison of several mammals. Nature 2005;434:338–345.
43 Laguette M-J, Abrahams Y, Prince S, Collins M: Sequence variants within the 3′-UTR of the *COL5A1* gene alters mRNA stability: implications for musculoskeletal soft tissue injuries. Matrix Biol J Int Soc Matrix Biol 2011;30:338–345.
44 Collins M, Posthumus M: Type V collagen genotype and exercise-related phenotype relationships: a novel hypothesis. Exerc Sport Sci Rev 2011;39:191–198.
45 Brown JC, Miller C-J, Posthumus M, Schwellnus MP, Collins M: The *COL5A1* gene, ultra-marathon running performance, and range of motion. Int J Sports Physiol Perform 2011;6:485–496.
46 Brown JC, Miller C-J, Schwellnus MP, Collins M: Range of motion measurements diverge with increasing age for *COL5A1* genotypes. Scand J Med Sci Sports 2011;21:e266–e272.
47 Collins M, Mokone GG, September AV, van der Merwe L, Schwellnus MP: The *COL5A1* genotype is associated with range of motion measurements. Scand J Med Sci Sports 2009;19:803–810.
48 Posthumus M, Schwellnus MP, Collins M: The *COL5A1* gene: a novel marker of endurance running performance. Med Sci Sports Exerc 2011;43:584–589.
49 Abrahams S, Posthumus M, Collins M: A polymorphism in a functional region of the *COL5A1* gene: association with ultraendurance-running performance and joint range of motion. Int J Sports Physiol Perform 2014;9:583–590.
50 O'Connell K, Posthumus M, Schwellnus MP, Collins M: Collagen genes and exercise-associated muscle cramping. Clin J Sport Med 2013;23:64–69.
51 Knapik JJ, Bauman CL, Jones BH, Harris JM, Vaughan L: Preseason strength and flexibility imbalances associated with athletic injuries in female collegiate athletes. Am J Sports Med 1991;19:76–81.
52 Knapik JJ, Jones BH, Bauman CL, Harris JM: Strength, flexibility and athletic injuries. Sports Med 1992;14:277–288.

53 Krivickas LS, Feinberg JH: Lower extremity injuries in college athletes: relation between ligamentous laxity and lower extremity muscle tightness. Arch Phys Med Rehabil 1996;77:1139–1143.
54 Hakim AJ, Cherkas LF, Grahame R, Spector TD, MacGregor AJ: The genetic epidemiology of joint hypermobility: a population study of female twins. Arthritis Rheum 2004;50:2640–2644.
55 Bertuzzi R, Pasqua LA, Bueno S, Lima-Silva AE, Matsuda M, Marquezini M, et al: Is the *COL5A1* rs12722 gene polymorphism associated with running economy? PloS One 2014;9:e106581.
56 Lim S-T, Kim C-S, Kim W-N, Min S-K: The *COL5A1* genotype is associated with range of motion. J Exerc Nutr Biochem 2015;19:49–53.
57 Tringali C, Brivio I, Stucchi B, Silvestri I, Scurati R, Michielon G, et al: Prevalence of a characteristic gene profile in high-level rhythmic gymnasts. J Sports Sci 2014;32:1409–1415.
58 Bell RD, Shultz SJ, Wideman L, Henrich VC: Collagen gene variants previously associated with anterior cruciate ligament injury risk are also associated with joint laxity. Sports Health Multidiscip Approach 2012;4:312–318.
59 Karsenty G, Park RW: Regulation of type I collagen genes expression. Int Rev Immunol 1995;12:177–185.
60 Jin H, van't Hof RJ, Albagha OME, Ralston SH: Promoter and intron 1 polymorphisms of *COL1A1* interact to regulate transcription and susceptibility to osteoporosis. Hum Mol Genet 2009;18:2729–2738.
61 Jin H, Evangelou E, Ioannidis JPA, Ralston SH: Polymorphisms in the 5′ flank of *COL1A1* gene and osteoporosis: meta-analysis of published studies. Osteoporos Int 2011;22:911–921.
62 Makareeva E, Cabral WA, Marini JC, Leikin S: Molecular mechanism of alpha 1(I)-osteogenesis imperfecta/Ehlers-Danlos syndrome: unfolding of an N-anchor domain at the N-terminal end of the type I collagen triple helix. J Biol Chem 2006;281:6463–6470.
63 Khoschnau S, Melhus H, Jacobson A, Rahme H, Bengtsson H, Ribom E, et al: Type I collagen alpha1 Sp1 polymorphism and the risk of cruciate ligament ruptures or shoulder dislocations. Am J Sports Med 2008;36:2432–2436.
64 Young BB, Zhang G, Koch M, Birk DE: The roles of types XII and XIV collagen in fibrillogenesis and matrix assembly in the developing cornea. J Cell Biochem 2002;87:208–220.
65 Oh SP, Taylor RW, Gerecke DR, Rochelle JM, Seldin MF, Olsen BR: The mouse alpha 1(XII) and human alpha 1(XII)-like collagen genes are localized on mouse chromosome 9 and human chromosome 6. Genomics 1992;14:225–231.
66 O'Connell K, Posthumus M, Collins M: No association between *COL3A1, COL6A1* or *COL12A1* gene variants and range of motion. J Sports Sci 2013;31:181–187.
67 Craib MW, Mitchell VA, Fields KB, Cooper TR, Hopewell R, Morgan DW: The association between flexibility and running economy in sub-elite male distance runners. Med Sci Sports Exerc 1996;28:737–743.
68 Jones AM: Running economy is negatively related to sit-and-reach test performance in international-standard distance runners. Int J Sports Med 2002;23:40–43.
69 O'Connell KS: Investigation of selected collagen genes in exercise-related musculoskeletal soft tissue phenotypes. 2014. https://open.uct.ac.za/handle/11427/12960 (accessed February 19, 2016).
70 O'Connell K, Posthumus M, Collins M: *COL6A1* gene and Ironman triathlon performance. Int J Sports Med 2011;32:896–901.
71 Banos CC, Thomas AH, Kuo CK: Collagen fibrillogenesis in tendon development: current models and regulation of fibril assembly. Birth Defects Res C Embryo Today 2008;84:228–244.
72 Liu SH, Yang RS, al-Shaikh R, Lane JM: Collagen in tendon, ligament, and bone healing. A current review. Clin Orthop Relat Res 1995;318:265–278.
73 Fleischmajer R, Perlish JS, Burgeson RE, Shaikh-Bahai F, Timpl R: Type I and type III collagen interactions during fibrillogenesis. Ann N Y Acad Sci 1990;580:161–175.
74 Olsen BR: New insights into the function of collagens from genetic analysis. Curr Opin Cell Biol 1995;7:720–727.
75 Lampe A, Bushby K: Collagen VI related muscle disorders. J Med Genet 2005;42:673–685.
76 Symoens S, Renard M, Bonod-Bidaud C, Syx D, Vaganay E, Malfait F, et al: Identification of binding partners interacting with the α_1-N-propeptide of type V collagen. Biochem J 2011;433:371–381.
77 Bertini E, Pepe G: Collagen type VI and related disorders: Bethlem myopathy and Ullrich scleroatonic muscular dystrophy. Eur J Paediatr Neurol 2002;6:193–198.
78 Okada M, Kawahara G, Noguchi S, Sugie K, Murayama K, Nonaka I, et al: Primary collagen VI deficiency is the second most common congenital muscular dystrophy in Japan. Neurology 2007;69:1035–1042.
79 Bonaldo P, Braghetta P, Zanetti M, Piccolo S, Volpin D, Bressan GM: Collagen VI deficiency induces early onset myopathy in the mouse: an animal model for Bethlem myopathy. Hum Mol Genet 1998;7:2135–2140.
80 Izu Y, Ansorge HL, Zhang G, Soslowsky LJ, Bonaldo P, Chu M-L, et al: Dysfunctional tendon collagen fibrillogenesis in collagen VI null mice. Matrix Biol J Int Soc Matrix Biol 2011;30:53–61.

81 Tanaka T, Ikari K, Furushima K, Okada A, Tanaka H, Furukawa K-I, et al: Genome-wide linkage and linkage disequilibrium analyses identify *COL6A1*, on chromosome 21, as the locus for ossification of the posterior longitudinal ligament of the spine. Am J Hum Genet 2003;73:812–822.
82 Schwellnus MP, Derman EW, Noakes TD: Aetiology of skeletal muscle 'cramps' during exercise: a novel hypothesis. J Sports Sci 1997;15:277–285.
83 Shang G, Collins M, Schwellnus MP: Factors associated with a self-reported history of exercise-associated muscle cramps in Ironman triathletes: a case-control study. Clin J Sport Med 2011;21:204–210.
84 Schwellnus MP: Cause of exercise-associated muscle cramps (EAMC) – altered neuromuscular control, dehydration or electrolyte depletion? Br J Sports Med 2009;43:401–408.
85 Schwellnus MP, Allie S, Derman W, Collins M: Increased running speed and pre-race muscle damage as risk factors for exercise-associated muscle cramps in a 56-km ultra-marathon: a prospective cohort study. Br J Sports Med 2011;45:1132–1136.
86 Kubo K, Yata H, Tsunoda N: Effect of gene polymorphisms on the mechanical properties of human tendon structures. Springerplus 2013;2:343.
87 Foster BP, Morse CI, Onambele GL, Williams AG: Human *COL5A1* rs12722 gene polymorphism and tendon properties in vivo in an asymptomatic population. Eur J Appl Physiol 2014;114:1393–1402.
88 Nishiyama T, McDonough AM, Bruns RR, Burgeson RE: Type XII and XIV collagens mediate interactions between banded collagen fibers in vitro and may modulate extracellular matrix deformability. J Biol Chem 1994;269:28193–28199.
89 Lapiere CM, Nusgens B, Pierard GE: Interaction between collagen type I and type III in conditioning bundles organization. Connect Tissue Res 1977;5:21–29.

Dr. Michael Posthumus
Division of Exercise Science and Sports Medicine
University of Cape Town
PO Box 115, Newlands 7925 (South Africa)
E-Mail michael.posthumus@uct.ac.za

Posthumus M, Collins M (eds): Genetics and Sports, ed 2, revised, extended.
Med Sport Sci. Basel, Karger, 2016, vol 61, pp 105–119 (DOI: 10.1159/000445244)

Genetic Testing for Sports Performance, Responses to Training and Injury Risk: Practical and Ethical Considerations

Alun G. Williams[a, b] • Henning Wackerhage[c] • Stephen H. Day[a]

[a]MMU Sports Genomics Laboratory, Manchester Metropolitan University, Crewe, [b]Institute of Sport, Exercise and Health, University College London, London, and [c]School of Medical Sciences, University of Aberdeen, Aberdeen, UK

Abstract

This paper addresses practical and ethical considerations regarding genetic tests to predict performance and/or risk of exercise-related injury or illness. Various people might wish to conduct sport related genetic tests for a variety of reasons. For example, an individual might seek personal genetic information to help guide their own sport participation. A sports coach might wish to test young athletes to aid team selection or individualize training. A physician might want to predict the risk of injury or illness in athletes and advise regarding selection or preventative measures. An insurance company might seek to estimate the risk of career-threatening injury for athletes based partly on genetic information. Whilst this information is, in part, encoded in our DNA sequence, the available tests allow generally only a poor prediction of the aforementioned variables. In other words, the current genetic tests and analysis methods are not powerful enough to inform important decisions in sport to a substantial degree. It is particularly disappointing that more than half of the commercially available genetic tests related to exercise and sport do not appear to identify publicly the genetic variants they assess, making scrutiny by academic scholars and consumers (or their representatives) impossible. There are also challenging ethical issues to consider. For example, the imposition of genetic tests on individuals (especially young people) by third parties is potentially susceptible to abuse. Scientists and practitioners should understand the limitations of the tests currently available, the ethical concerns and the importance of counselling before and after testing so that they are only used in a responsible manner.

The focus of this paper is on the potential use of genetic tests to predict performance and/or the risk of exercise-related injury or illness. Several earlier papers of this book, including those of Venezia and Roth (regarding genes and training adaptations), Ahmetov et al. (regarding genes and sport performance), Rahim et al. (regarding genes and musculoskeletal injuries) and Collins et al. (genes and musculoskeletal phenotypes) [this vol., pp. 29–40, 41–54, 68–91, 92–104], contain information and ideas relevant to the notion of testing athletes or other individuals for sport- and exercise-related traits. However, this review will deliberately focus on both the scientific and ethical issues regarding the validity, utility and practicality of genetic testing of athletes at the current time.

Commercially Available Tests

There are already several commercial operations that offer 'mail order' or 'direct-to-consumer' (DTC) genetic testing of polymorphisms apparently associated with physical performance capacity or exercise-related health. We believe the first commercial operation began in Australia in 2004 and was marketed as the ACTN3 Sports Performance Test™ by the company Genetic Technologies. Most recently, surveys [1, 2] of available tests identified 39 companies providing DTC genetic tests marketed in relation to human sport or exercise performance or injury (table 1).

In that recent analysis mentioned [1], the most commonly tested variant remained the R577X variant in the *ACTN3* gene, tested by 89% of the 18 companies that appeared to present information about their genetic tests on their websites. That observation corresponds with assessments that *ACTN3* R577X is currently the polymorphism with the strongest scientific evidence in support of an association with athlete phenotypes [3–5]. However, there is limited information that can be gleaned from discrete, single-marker genetic tests at common polymorphisms, beyond an 'interest' at an individual level. Therefore, the companies that test only a single variant should clearly not claim to provide information on which personal exercise training or sport decisions can reasonably be made. The level of qualification and explanation given alongside the raw genetic information to individuals also varied considerably, as pointed out previously [6]. Some companies appear to treat the genetic data in a cautious manner and are suitably careful not to extend preliminary scientific findings into claims that extend beyond the published scientific literature base. However, that sensible approach is not universally adopted, and thus some of the claims (overt or implied) for the extent of the usefulness of the single genetic marker information are not supported by sufficiently strong scientific evidence. Thus, some individuals might make decisions about their personal exercise and sport participation on the basis of DTC genetic test information that are not warranted. It is therefore understandable to some extent why the majority of the companies identified [1] as offering defined DTC genetic tests assessed a panel of multiple genetic variants. The next most tested variant was in the *ACE* gene (61% of the 18 companies), while the median number of variants tested was 6, ranging from 1 to 27. However, as one considers genetic variants beyond those in the *ACTN3* and *ACE* genes that are reasonably well studied, the level of scientific evidence to support the choice of any particular polymorphism reduces

Table 1. Companies found to be providing DTC genetic tests marketed as being related to sport and exercise performance or sports injury

Company	Website name	Variants tested, n	Genes of variants tested (according to websites/client reports)
23andMe	23andme.com	1	*ACTN3*
Advanced Business Services	abservices.eu	n.f.	n.f.
Agoga	agoga.com.au	n.f.	n.f.
Asper Biotech	asperbio.com	2	*ACE, ACTN3*
Athletigen	athletigen.com	18	*ACTN3, ADRB2, AGT, AMPD1, CILP, COL1A1, COL5A1, CREB1, IL1B, HIF1A, MMP3, NAT2, PPARD, PPARG, PPARGC1A, RAD23A*
Atlas Sports Genetics	atlasgene.com	1	*ACTN3*
C2DNA	c2dna.com	n.f.	n.f.
Cosmetics DNA	cosmetics-dna.com	n.f.	n.f.
CyGene Direct	cygene.infinityarts.com	6	*ACE, APOE, BDKRB2, ENOS, VDR*
DNA Fit	dnafit.com	21	*ACE, ACTN3, ADRB2, AGT, BDRKB2, COL1A1, COL5A1, CRP, GDF5, GSTM1, GSTT1, IL6, IL-6R, NRF-2, PPARA, PPARGC1A, SOD2, TNF, TRHR, VDR, VEGF*
DNA Spectrum	dnaspectrum.com	7	*ACE, ACTN3, ADRB2, ADRB3, ENOS, FTO*
DNAeX	dnaex.net	14	*ACE, ACTN3, ADRB2, AGT, BDKRB2, COL5A1, CRP, IL6, NRF, PPARA, PPARGC1A, TRHR, VDR, VEGF*
DNAlysis	dnalysis.co.za	19	*ACE, ACTN3, ADRB2, AGT, BDRKB, COL1A1, COL5A1, CRP, GDF5, IL6, IL-6R, NRF-2, PPARA, PPARGC1A, SOD2, TNF, TRHR, VDR, VEGF*
GenePlanet	geneplanet.com	2	*ACTN3, PPARA*
Genetic Center	genetic-center.com	n.f.	n.f.
Genetic Performance	geneticperformance.com	n.f.	n.f.
Genetrainer	genetrainer.com	n.f.	n.f.
Gknowmix	gknowmix.com	n.f.	n.f.
Gonidio	gonidio.com	27	*ACE, ACTN3, ADRA2A, ADRB1, ADRB2, AMPD1, BDKRB2, CHRM2, CK-MM, COL1A1, COL5A1, DIO1, EPOR, HBB, HIF-1, MCT-1, MMP3, NOS3, PPARD, PPARGC1, VDR, VEGF*
IgnitePlay	igniteplay.com	n.f.	n.f.
Institute for Optimum Nutrition	ion.ac.uk	n.f.	n.f.
Lyfgene DNA	lyfgene.com	n.f.	n.f.

Table 1. Continued

Company	Website name	Variants tested, n	Genes of variants tested (according to websites/client reports)
Molecular Testing Labs Fitness	mtlfitness.com	9	*ACTN3, EDN1, INSIG2, LIPC, LPL, MMP3, PPARD, PPARGC1A, SLC30A9*
Musclegenes	musclegenes.com	n.f.	n.f.
MyInnerGo	myinnergo.com	n.f.	n.f.
MyRISQ	myrisq.com	n.f.	n.f.
Nordic Laboratories	nordiclabs.com	20	*ACE, ACTN3, ADRB2, AGT, BDRKB, COL1A1, COL5A1, CRP, GDF5, IL6, IL6R, NRF, PPARA, PPARGC1A, SOD2, TNF, TRHR, VDR, VEGF*
Nutragene	nutragene.com	n.f.	n.f.
Pathway Genomics	pathway.com	n.f.	n.f.
Phenom Biosciences	iamaphenom.com	n.f.	n.f.
PlayDNA	playdna.co.uk	n.f.	n.f.
Simplified Genetics	simplifiedgenetics.com	5	*ADRB2, ADRB3, APOE, PPARG*
Sports Gene	sportsgene.ee	6	*ACE, ACTN3, AMPD1, GDF8, NOS3, PPARGC1A*
The Wellness Gene	wellnessgene.com	23	*ACE, ACTN3, AMPD1, BDKRB2, CHRM2, CKMM, COL1A1, COL5A1, DIO1, EPOR, HBB, HIF1A, MCT1, MMP3, NOS3, PPARD, PPARGC1A, VDR, VEGF*
TheMakingsofMe	themakingsofme.com	3	*ACTN3, HIF1A, NOS3*
ThinnerGene	thinnergene.com	n.f.	n.f.
ThriveLondon	thrivelondon.com	n.f.	n.f.
Woblab	woblab.com	6	*ACE, ACTN3, AMPD1, GDF8, NOS3, PPARGC1A*
XRGenomics	xrgenomics.com	n.f.	n.f.

Data may not be 100% accurate because it is dependent on the ability to navigate the websites appropriately, and updating of the information provided on the websites. Gene names are in most instances listed verbatim as presented on the company websites, even though some gene names given did not conform to the standard nomenclature. n.f. = Information not found. From the supplementary file in Webborn et al. [1], with permission.

considerably in volume, and the scientific evidence is considered weak by the majority of sport and exercise genetics researchers [1, 4, 7, 8], including ourselves. While commercial pressures undoubtedly exist, it would be wise, and more responsible, to wait for a greater scientific consensus before offering tests that currently have only weak supporting evidence. Counselling that puts the genetic information – including the limitations of its usefulness – into the proper context is recommended as a minimum, although not even a sophisticated counselling service can resolve scientific controversy.

It is particularly disappointing that 54% of the companies offering DTC genetic tests related to exercise and sport did apparently not publicly state which genetic variants they assessed [1]. Again, while commercial pressures undoubtedly exist, it is impossible for anyone – academic scholar or otherwise – to scrutinize the service provided by the companies if the detail is not presented to consumers. The detail is absolutely crucial, because quite literally millions of genetic tests could theoretically be conducted, and the choice of variants to be tested – and how those results are interpreted – is fundamental to the usefulness of the test. Such apparent secrecy is presumably due to commercial sensitivity in part, although failing to publicize the tests conducted is arguably a tacit admission that the scientific evidence supporting the genetic variants chosen is weak. Perhaps the specific genetic variants tested by a particular company will change over time as scientific knowledge in this field progresses, but if that happens, then it severely questions the validity of the original test.

In broad terms, based on scientific evidence, the information provided by these tests may be of interest to many people and may help individuals (or sports coaches, etc.) attempt to 'better understand' their observed physical limitations to performance or training adaptations. However, there is currently little evidence (there is lack of replication, in particular [9]) that these kinds of tests provide information regarding either predisposition for a particular sport or prediction of the training response likely to occur to a particular training programme that are useful in a practical sense. For example, a thorough multidisciplinary analysis of the efficacy of these tests in talent identification would need to be conducted to determine whether genetic data provide information not already captured within other, traditional non-genetic tests of physiological, anthropometric and performance characteristics that are already routinely used in talent identification.

Nevertheless, the availability of such tests for virtually anyone to access is certainly an interesting development and probably heralds a new era of genetic testing for all manners of individual characteristics. It is likely that such tests will become more widely requested, at least in part, because recent legislation in the USA ensures that consumers need not fear insurance or employment discrimination on the basis of genetic test results [10]. However, a significant debate exists regarding the science that is purported to underpin such tests. In addition, there are wider ethical concerns regarding issues such as the confidentiality of data, the need for counselling when interpreting personal genetic data, the use of genetic data in assessing insurance risk, etc. We will now address such issues in this paper.

Is There a Fundamental Difference between Genetic Tests and Non-Genetic Tests?

Discrimination on the basis of partially inherited factors is already common in sport today – for example, when strength is measured to identify children that have a talent for weightlifting or to select a national weightlifting team. So does it matter whether talent is identified by genetic or by non-genetic tests? Is there a fundamental difference?

To give an example, one could rightly ask: 'What is the difference between a performance test that measures a variable which predicts ~2% of an individual's muscle power as opposed to the commercially available *ACTN3* R577X genetic test which perhaps measures something very similar?' The term 'genetic exceptionalism' is sometimes used in this debate meaning that genetic tests are special and thus require specific legislation that is different from that for other biological tests [11]. In their analysis, Green and Botkin [11] focus on medical tests and conclude that 'no clear, significant distinctions between genetic and non-genetic tests justify a different approach [...]' although they state that genetic tests may lead to stigmatization, family discord and psychological stress – but much of this also applies to some non-genetic tests such as HIV tests.

The working party that developed the BASES position stand in *Genetic Research and Testing in Sport and Exercise Science* [12] identified 2 fundamental differences between genetic and non-genetic (performance) tests, and we include a third in this paper as we did previously [13]:

(1) The information gained from a genetic test does not change with age, whereas the information derived from a traditional performance test, and consequently the predictive quality of that test, does change with age

(2) Not all the predictions that can be made with a genetic test may be known at the time when the genetic test is conducted

(3) Genetic tests have more implications for relatives and partners than other tests

These three points are now discussed in more detail:

(1) The genome of an individual and the variations therein are constant throughout life (there are exceptions not relevant for this argument). Consequently, genetic tests can be performed as soon as the DNA of an individual can be obtained – even before birth. This is fundamentally different from other tests such as a lactate test where the information derived from the test, and consequently its predictive quality (for example, the ability to predict marathon running ability) depends very much on the age when this test is performed. This difference is important because it opens up possibilities for the misuse of genetic tests, especially when used before birth or in minors. One scenario could be that these tests are used to select embryos on the basis of sporting, intellectual or other abilities (a 'new eugenics'?). Also, with young children such tests could be used to make decisions for or against a sporting career.

(2) Both genetic and non-genetic performance tests may later be linked to other variables such as disease. We think that this risk is higher for genetic tests than for other performance tests. However, common performance-related variables have also been linked to disease. For example, muscle strength has been inversely associated with mortality [14], which means a hitherto unanticipated higher risk of death for individuals who had previously performed poorly in a strength test. However, our opinion is that the risk of an unanticipated disease link, especially to a specific disease, is

probably higher for genetic tests than for other biomedical tests. One past example is a polymorphism in the *APOE* gene (which encodes a protein important for lipid transport). Based on 2 loci, there are 3 polymorphic forms: APO E2, E3 and E4. The APO E4 variant was initially shown to be associated with limited differences in lipid profile but only later with late-onset familial Alzheimer's disease [15]. Thus, individuals who had previously been tested for a genetic variant that was associated with their lipid profile now retrospectively had information about their likelihood of suffering from the severer Alzheimer's disease. How can we deal with the risk of unanticipated future links of genetic test results to severe disease? A pragmatic solution would be to make mandatory genetic counselling necessary so that individuals are aware of this risk before deciding whether to proceed or not with a genetic test.

(3) A third difference is that genetic tests have more direct implications for relatives than other tests, as Green and Botkin [11] point out. Some non-genetic tests have important implications for others – for example, a positive HIV test result may not only affect the individual, but also have far-reaching implications for relatives, plus current and previous sexual partners. However, genetic tests are *always* also predictive for close relatives and this should be taken into account when performing genetic tests. Again, mandatory genetic counselling could be useful to at least ensure that individuals understand the implications of genetic test results for their relatives.

The first difference we have highlighted in particular, and its resultant consequences such as the potential selection of embryos, warrants the use of the term 'genetic exceptionalism'. Also, for now at least, scientists must accept that 'people see genetic information as special' [16], even if in reality the information obtained by genetic and non-genetic tests can in some instances be very similar.

How Powerful Are the Genetic Tests That Currently Exist?

By the end of 2007, there had been over 200 genetic variations associated with performance and health-related fitness phenotypes [17], and that number continues to increase although it has not been documented in the same comprehensive way in recent years. However, in many cases, there has still only been a single positive association with a relevant phenotype, and consistent replication of the reported associations is obviously needed to increase confidence in the associations.

Focusing initially on rare mutations, we are aware of just two that appear to have a powerful 'positive' influence on exercise-related phenotypes (although many rare mutations exist that have powerful 'negative' influences on exercise-related phenotypes). One rare mutation exists in the family of the Finnish cross-country skier Eero Antero Mäntyranta, where over 200 g of haemoglobin per litre of blood and extremely high haematocrit values of >60% were reported [18]. Erythropoietin (EPO) levels were normal but the mutation enhanced sensitivity to EPO. The responsible mutation in the EPO receptor (*EPOR*) gene was subsequently identified [19] as a premature stop codon that shortens the protein by 70 amino acids. A more active EPOR protein is the result and is likely to have contributed to Mäntyranta winning 3 gold and 4 other medals at the Olympic Games during the 1960s. Haematocrit in excess of 50% would attract investigation from the antidoping authorities, although the WADA 'Athlete Biological Passport' concept combined with DNA analysis may allow an athlete competing today with such a mutation to demonstrate that no erythropoiesis-stimulating agents had been used.

A second naturally occurring rare mutation, this time with profound effects on muscle growth, has been observed in the myostatin gene (*MSTN* [20]). The authors reported the existence of a boy who was 'extraordinarily muscular, with protruding muscles in his thighs and upper arm'. The boy was homozygous for an extremely rare G-to-A intronic mutation in the *MSTN* gene and his muscle tissue showed no evidence of functional MSTN protein. This observation in a human parallels the deliberate knockout of the myostatin gene in mice, which has resulted in pronounced skeletal muscle fibre growth [21]. However, it is not certain that health status (e.g. cardiac function) will not be compromised by the absence of functional MSTN protein, and there may be other detrimental changes such as changes in tendon properties [22]. Nevertheless, one possible kind of genetic test for physical performance could be to screen athletes for rare mutations such as those described in the *EPOR* and *MSTN* genes. The rarity and heterogeneity of such powerful mutations means that they would need to be identified using sequencing technology that allows base-by-base examination of relevant chromosomal regions, and even then would probably have a very low success or 'hit' rate. Nevertheless, transgenic mouse models suggest that there are multiple genes whose mutation can have large effects on performance [23], and it seems likely that some of these mutations occur in some humans. Thus, more athletes with rare, powerful mutations will probably be discovered. This information might help to explain the physiological capacity of those individuals, provide clues regarding the type of events in which they might excel and even provide information that might be useful in the individualization of training.

In contrast to rare, powerful mutations, there are many common polymorphisms that are usually less powerfully associated with performance-related phenotypes. Here, we will consider as an example just *ACTN3* R577X, while others are addressed in other reviews.

A common polymorphism exists in the *ACTN3* gene in which the element encoding arginine (R) may alternatively be a stop codon (X) that produces a truncated protein molecule with no known function. Studies of elite Australian [24] and Finnish [25] sprint/power athletes found none were of the XX genotype. An *ACTN3* knockout mouse displaying characteristics summarized as a fast-to-slow muscle phenotype change helps to explain mechanistic links between *ACTN3* genotype and physical performance phenotype [26, 27]. These data suggest it is virtually impossible to become an elite sprint or power athlete without the R allele of *ACTN3*, suggesting a powerful influence of the *ACTN3* common polymorphism. However, in the general population it appears that only at most ~2% of interindividual variability in muscle strength, sprinting speed and competitive racing distances are related to the *ACTN3* genotype [28–30], and this seems somewhat at odds with the striking absence of the XX genotype in most elite sprint/power athletes. This paradox may be analogous with a classic study demonstrating that while the maximal rate of oxygen uptake may have good statistical power in predicting endurance running performance amongst a wide range of running abilities, when a narrow (elite) range of performance is considered it is then running economy, a different phenotype, that becomes the dominant factor influencing performance [31]. Perhaps the *ACTN3* genotype (and thus the presence/absence of α-actinin-3 protein in muscle) is equivalent to running economy in that analogy, in that only amongst highly trained athletes does the practical significance of the *ACTN3* genotype increase to a meaningful extent. Nevertheless, even then, the *ACTN3* XX genotype does not seem to completely preclude success in power events [32]. So the practical use of an *ACTN3* genotype test to inform an individual (or his/her coach, for example) about future performance potential is thus still not clear, although of all common polymorphisms *ACTN3* currently shows the most potential in this regard. Despite scientific uncertainty, one can understand individuals interested in exercise and sport wishing to learn about their own genetic composition at this locus – even if this discrete variant only imparts a very small proportion of the total genetic influence. So the provision of a service for the testing of the *ACTN3* R577X polymorphism on a commercial basis could be seen as meeting an understandable public interest and providing information that has at least some replicated scientific evidence to justify the activity. Nevertheless, the predictive value of such tests in the context of training responses or talent identification in sport is virtually zero [1, 8].

It is our view that while *ACTN3* and other genotypes show promise, a large monogenic influence of a common polymorphism on physical performance or injury phenotypes is extremely unlikely. Future genetic testing of athletes will therefore need to account for a greater proportion of the genetic influence on interindividual variability via simultaneous consideration of genetic variation at a number of loci. The principle behind this polygenic approach has been demonstrated [33], although more research is needed before the application of genetic algorithms to athletes can proceed with a valid scientific basis. Major decision-making regarding the careers or training/medical environment of athletes, based largely on the genetic testing of those athletes, is not scientifically justified until more research is conducted and therefore would be unethical at the present time. An increased predictive power of genetic testing in athletes will be dependent upon the quality of the data that emerge from the future research that will itself be dependent on the application of more advanced technology such as cheap, fast, whole-genome sequencing.

It is therefore doubtful whether the body of knowledge is sufficiently strong for practical applications such as selection of talented athletes for intensive training, individualization of training regimens to improve performance or modulation of the training load to minimize injury risk. Perhaps at the most elite level of sports performance, where every little piece of information about an athlete is most likely to be considered sensibly in the context of much other data, some tentative conclusions might be reached and any implementation evaluated extremely closely. However, the careful wording of the previous sentence is very deliberate, and there is a requirement for greater replication of the >250 genotype-phenotype associations reported to date [9, 17, 34] before genetic testing in sport and exercise has more widespread utility. This is because the common polymorphisms identified to date only account, individually, for a small proportion of the interindividual variability in phenotype. To explain a larger proportion of the variability, either rare variants of large effect or favourable combinations of many common variants need to be identified. Evidence regarding rare variants of large effect is currently limited to very few documented examples [19, 20]. Using several common variants, elite athletes in certain sports have at least been shown to differ in polygenic profile from non-athletes and from elite athletes in other sports [35–38], and such differences will become clearer as broader panels of appropriate variants are included [33]. Quite recently, 21 single-nucleotide polymorphisms were identified that appear to capture the heritable component (approx. 50% of total interindividual variability) of the responsiveness to endurance training of the maximal rate of oxygen uptake phenotype [39]. While this observation also needs replication (and that is more easily written than conducted in practice), the applications of this kind of insight into an individual's potential to respond to training in a sporting context, as well as an exercise-health-fitness context, are obvious.

Who Should Be Able to Request Genetic Tests?

With an accumulating number of available genetic tests, a wide range of genetic testing scenarios seems inevitable in the sport and exercise context. The 5 examples below show different individuals (athletes, parents, coaches, physicians) requesting genetic tests either for themselves or for others for a variety of purposes:

(1) Genetic performance testing: parents might request genetic tests for genes that determine adult peak maximal oxygen uptake for their child in order to decide whether to send their offspring to a special sports school or not

(2) Genetic testing for personalized training programmes: a coach might request genetic tests for trainability polymorphisms in order to decide whether weightlifters should take part in a high-volume or high-intensity strength training programme (see Roth [40] for a more detailed debate)

(3) Genetic sports disease testing: sports associations and their physicians might use genetic tests to screen for variants associated with sudden death in order to identify individuals that are at high risk and to exclude such individuals from competing. A similar but non-genetic screening programme is performed in Italy [41] and has

reduced sudden deaths but has also led to the exclusion of elite athletes from competitions

(4) Genetic sports injury testing: an individual could request a DTC genetic test for tendon gene variants to predict the frequency of injury in order to decide whether to embark on a career as a professional football player or not

(5) Genetic insurance risk testing: an insurance company might request genetic tests for injury genes in order to determine the insurance premium for a professional football player

At the moment, most people would probably accept that adults should not be prohibited from requesting genetic tests for performance or sport and health-related traits for themselves, in order to assist making life choices. We recommend, however, even in the case of DTC tests [42] such as those listed in table 1, that these tests are accompanied by some genetic counselling. This could include counselling about potential unanticipated associations with disease, implications for relatives and other information which has been summarized in a review on 'ideal' genetic counselling [43]. The other recommendation is that genetic tests should, for the time being at least, only be allowed to be requested by mature individuals (in the sense of mental capacity, in specific relation to the issue of genetic testing in sport) who understand the implications of a test.

Serious ethical concerns arise if genetic tests for an individual are requested by others. In the sport context this could be physicians, parents or coaches, and such tests could be used to discriminate in favour of or against athletes on the basis of performance-related or health-related genetic information. Some of these concerns are highlighted by a real case [44]: Eddy Curry was a professional NBA basketball player who missed games due to cardiovascular symptoms. His team (the Chicago Bulls) requested a genetic test for him based on the advice of a cardiologist. The athlete refused to do the test and was sold to the New York Knicks who did not request such a test. This is an example of an employer trying to force an employee to do a genetic test in order to then discriminate for or against the employee based on the test result. In this case one can see the team's perspective as an employer that has a duty of care that might want to prepare itself for an acute cardiovascular problem, yet on the other hand one can understand the athlete's desire not to know or have others know about a serious genetic condition.

The most serious ethical concern arises from the major fundamental difference between genetic and non-genetic testing which is that the DNA does not change throughout life and that therefore prenatal genetic tests could be performed to select an embryo with the 'best sport genotype' or to abort an embryo with the 'wrong sport genotype'. The legal prevention of such testing is important because there is evidence that some parents use prenatal information to decide about whether or not to abort a pregnancy. In India and China, non-genetic ultrasound scans have been used to determine the sex of an embryo, and this practice has greatly contributed to an estimated 80 million 'missing' females [45]. Similarly, it may be that some genetic tests (not limited to genetic sex tests), including genetic tests of sporting potential, would be used by some parents during pre-implantation genetic diagnosis or to decide whether or not to select an embryo instead of others.

Conclusions

The focus of this paper has been on the potential use of genetic tests to predict performance and/or the risk of exercise-related injury or illness. The knowledge base is expected to develop so that the prescription of training, nutrition and competition load, and the management of injury risk, can be conducted in a more individualized manner than is currently possible to improve both performance and athlete welfare [46]. Consequently, various people may wish to conduct a sport-related genetic test on themselves, or on another person, for a variety of reasons. An individual may seek personal genetic information to assist him/her with their own sporting participation and career, by identifying the most suitable type of sport. A sports coach may wish to test the members of a youth team to assist in selection for a professional career or to individualize training. A physician may want to predict the risk of injury or illness in an athlete and advise a coach regarding selection or preventative measures. An insurance company may seek to estimate the risk of career-threatening injury or illness to an athlete based partly on genetic information. However, despite the commercial availability of genetic tests today, the evidence available at present suggests that few, and probably none, of these or similar scenarios are scientifically justified – the genetic tests available at the moment are not powerful enough to provide valid data on which to base important decisions in sport.

As written recently [47], as the evidence base improves, we predict that the first evidence-based practical application of genomic information in the management of elite athletes will involve the modification of training in an attempt to reduce the likelihood of injury when an athlete is predicted from their genome to be at greater than average risk. Secondly, personalization of training programmes should become possible in sports where 2 or more dichotomous physiological characteristics combine to produce athletic performance. For example, in many team sports where both endurance and sprint ability are required, at a truly elite level (where physiological monitoring as well as strength and conditioning processes are of the highest standard) it should become possible to carefully modify the relative emphasis during training of the development of those two physiological characteristics according to genotype. Thirdly, and probably more distantly in the future, some selection processes in sport might become informed by genomic information, although even in those cases the genomic information will probably remain secondary to the more informative phenotypic information.

There are many challenging ethical issues regarding the genetic testing of athletes. The imposition of genetic tests on individuals by third parties, and particularly the imposition of genetic tests on young people, is potentially susceptible to abuse. We suggest that an individual must be able to consent to their own genetic test (i.e. not of another individual/employee/child), and this is only valid if he/she can demonstrate beforehand an understanding of the possible implications of the test result. We also recommend that an appropriate standard of genetic counselling is always provided. It is critical that these issues continue to be debated widely so that the tests already available, and the more powerful ones that are likely to emerge following improvements in knowledge and genetic technology, are only used in acceptable ways.

References

1 Webborn N, Williams A, McNamee M, Bouchard C, Pitsiladis Y, Ahmetov I, Ashley E, Byrne N, Camporesi S, Collins M, Dijkstra P, Eynon N, Fuku N, Garton FC, Hoppe N, Holm S, Kaye J, Klissouras V, Lucia A, Maase K, Moran C, North KN, Pigozzi F, Wang G: Direct-to-consumer genetic testing for predicting sports performance and talent identification: consensus statement. Br J Sports Med 2015;49:1486–1491.

2 Williams AG, Heffernan SM, Day SH: Genetic testing in exercise and sport – have direct-to-consumer genetic tests come of age? Sci Sport Curr Trends 2014;2:3–9.

3 Ma F, Yang Y, Li X, Zhou F, Gao C, Li M, Gao L: The association of sport performance with *ACE* and *ACTN3* genetic polymorphisms: a systematic review and meta-analysis. PLoS One 2013;8:e54685.

4 Eynon N, Hanson ED, Lucia A, Houweling PJ, Garton F, North KN, Bishop DJ: Genes for elite power and sprint performance: *ACTN3* leads the way. Sports Med 2013;43:803–817.

5 Heffernan SM, Kilduff LP, Erskine RM, Day SH, McPhee JS, McMahon GE, Stebbings GK, Neale JP, Lockey SJ, Ribbans WJ, Cook CJ, Vance B, Raleigh SM, Roberts C, Bennett MA, Wang G, Collins M, Pitsiladis YP, Williams AG: Association of *ACTN3* R577X but not *ACE* I/D gene variants with elite rugby union player status and playing position. Physiol Genomics 2016;48:196–201.

6 Wagner JK, Royal CD: Field of genes: an investigation of sports-related genetic testing. J Pers Med 2012;2:119–137.

7 Bouchard C: Overcoming barriers to progress in exercise genomics. Exerc Sport Sci Rev 2011;39:212–217.

8 Wang G, Padmanabhan S, Wolfarth B, Fuku N, Lucia A, Ahmetov, II, Cieszczyk P, Collins M, Eynon N, Klissouras V, Williams A, Pitsiladis Y: Genomics of elite sporting performance: what little we know and necessary advances. Adv Genet 2013;84:123–149.

9 Hughes DC, Day SH, Ahmetov II, Williams AG: Genetics of muscle strength and power: polygenic profile similarity limits skeletal muscle performance. J Sports Sci 2011;29:1425–1434.

10 National Human Genome Research Institute: Genetic information nondiscrimination act: 2007–2008. National Human Genome Research Institute 2008. http://www.genome.gov/24519851 (accessed May 10, 2015).

11 Green MJ, Botkin JR: 'Genetic exceptionalism' in medicine: clarifying the differences between genetic and nongenetic tests. Ann Intern Med 2003;138:571–575.

12 Williams AG, Wackerhage H, Miah A, Harris RC, Montgomery HE: Genetic Research and Testing in Sport and Exercise Science: BASES Position Stand. Leeds, British Association of Sport and Exercise Sciences, 2007.

13 Williams AG, Wackerhage H: Genetic testing of athletes; in Collins M (ed): Genetics and Sports. Med Sport Sci. Basel, Karger, 2009, vol 54, pp 176–186.

14 Ruiz JR, Sui X, Lobelo F, Morrow JR Jr, Jackson AW, Sjostrom M, Blair SN: Association between muscular strength and mortality in men: prospective cohort study. BMJ 2008;337:a439.

15 Strittmatter WJ, Saunders AM, Schmechel D, Pericak-Vance M, Enghild J, Salvesen GS, Roses AD: Apolipoprotein E: high-avidity binding to beta-amyloid and increased frequency of type 4 allele in late-onset familial Alzheimer disease. Proc Natl Acad Sci U S A 1993;90:1977–1981.

16 Human Genetics Commission: Genetic Information, Public Consultation: Second Annual Report of the Human Genetics Commission. London, Human Genetics Commission, 2002.

17 Bray MS, Hagberg JM, Perusse L, Rankinen T, Roth SM, Wolfarth B, Bouchard C: The human gene map for performance and health-related fitness phenotypes: the 2006–2007 update. Med Sci Sports Exerc 2009;41:35–73.

18 Juvonen E, Ikkala E, Fyhrquist F, Ruutu T: Autosomal dominant erythrocytosis caused by increased sensitivity to erythropoietin. Blood 1991;78:3066–3069.
19 De la Chapelle A, Traskelin AL, Juvonen E: Truncated erythropoietin receptor causes dominantly inherited benign human erythrocytosis. Proc Natl Acad Sci U S A 1993;90:4495–4499.
20 Schuelke M, Wagner KR, Stolz LE, Hubner C, Riebel T, Komen W, Braun T, Tobin JF, Lee S-J: Myostatin mutation associated with gross muscle hypertrophy in a child. N Engl J Med 2004;350:2682–2688.
21 McPherron AC, Lawler AM, Lee SJ: Regulation of skeletal muscle mass in mice by a new TGF-beta superfamily member. Nature 1997;387:83–90.
22 Mendias CL, Bakhurin KI, Faulkner JA: Tendons of myostatin-deficient mice are small, brittle, and hypocellular. Proc Natl Acad Sci U S A 2008;105:388–393.
23 Wackerhage H: Molecular Exercise Physiology: An Introduction. New York, Routledge, 2014.
24 Yang N, MacArthur DG, Gulbin JP, Hahn AG, Beggs AH, Easteal S, North K: *ACTN3* genotype is associated with human elite athletic performance. Am J Hum Genet 2003;73:627–631.
25 Niemi AK, Majamaa K: Mitochondrial DNA and *ACTN3* genotypes in Finnish elite endurance and sprint athletes. Eur J Hum Genet 2005;13:965–969.
26 MacArthur DG, Seto JT, Chan S, Quinlan KG, Raftery JM, Turner N, Nicholson MD, Kee AJ, Hardeman EC, Gunning PW, Cooney GJ, Head SI, Yang N, North KN: An *Actn3* knockout mouse provides mechanistic insights into the association between alpha-actinin-3 deficiency and human athletic performance. Hum Mol Genet 2008;17:1076–1086.
27 Seto JT, Quinlan KG, Lek M, Zheng XF, Garton F, MacArthur DG, Hogarth MW, Houweling PJ, Gregorevic P, Turner N, Cooney GJ, Yang N, North KN: *ACTN3* genotype influences muscle performance through the regulation of calcineurin signaling. J Clin Invest 2013;123:4255–4263.
28 Ahmetov II, Druzhevskaya AM, Lyubaeva EV, Popov DV, Vinogradova OL, Williams AG: The dependence of preferred competitive racing distance on muscle fibre type composition and *ACTN3* genotype in speed skaters. Exp Physiol 2011;96:1302–1310.
29 Clarkson PM, Devaney JM, Gordish-Dressman H, Thompson PD, Hubal MJ, Urso M, Price TB, Angelopoulos TJ, Gordon PM, Moyna NM, Pescatello LS, Visich PS, Zoeller RF, Seip RL, Hoffman EP: *ACTN3* genotype is associated with increases in muscle strength in response to resistance training in women. J Appl Physiol 2005;99:154–163.
30 Moran CN, Yang N, Bailey ME, Tsiokanos A, Jamurtas A, Macarthur DG, North K, Pitsiladis YP, Wilson RH: Association analysis of the ACTN3 R577X polymorphism and complex quantitative body composition and performance phenotypes in adolescent Greeks. Eur J Hum Genet 2007;15:88–93.
31 Conley DL, Krahenbuhl GS: Running economy and distance running performance of highly trained athletes. Med Sci Sports Exerc 1980;12:357–360.
32 Lucia A, Olivan J, Gomez-Gallego F, Santiago C, Montil M, Foster C: Citius and longius (faster and longer) with no alpha-actinin-3 in skeletal muscles? Br J Sports Med 2007;41:616–617.
33 Williams AG, Folland JP: Similarity of polygenic profiles limits the potential for elite human physical performance. J Physiol 2008;586:113–121.
34 Ahmetov II, Fedotovskaya ON: Sports genomics: current state of knowledge and future directions. Cell Mol Exerc Physiol 2012;1:e1.
35 Drozdovska SB, Dosenko VE, Ahmetov, II, Ilyin VN: The association of gene polymorphisms with athlete status in Ukrainians. Biol Sport 2013;30:163–167.
36 Ahmetov II, Williams AG, Popov DV, Lyubaeva EV, Hakimullina AM, Fedotovskaya ON, Mozhayskaya IA, Vinogradova OL, Astratenkova IV, Montgomery HE, Rogozkin VA: The combined impact of metabolic gene polymorphisms on elite endurance athlete status and related phenotypes. Hum Genet 2009;126: 751–761.
37 Ruiz JR, Arteta D, Buxens A, Artieda M, Gomez-Gallego F, Santiago C, Yvert T, Moran M, Lucia A: Can we identify a power-oriented polygenic profile? J Appl Physiol 2010;108:561–566.
38 Ruiz JR, Gomez-Gallego F, Santiago C, Gonzalez-Freire M, Verde Z, Foster C, Lucia A: Is there an optimum endurance polygenic profile? J Physiol 2009; 587:1527–1534.
39 Bouchard C, Sarzynski MA, Rice TK, Kraus WE, Church TS, Sung YJ, Rao DC, Rankinen T: Genomic predictors of maximal oxygen uptake response to standardized exercise training programs. J Appl Physiol 2011;110:1160–1170.
40 Roth SM: Perspective on the future use of genomics in exercise prescription. J Appl Physiol 2008;104: 1243–1245.
41 Corrado D, Migliore F, Zorzi A, Siciliano M, Basso C, Schiavon M, Thiene G: Preparticipation electrocardiographic screening for the prevention of sudden death in sports medicine (in Italian). G Ital Cardiol 2011;12:697–706.
42 Hogarth S, Javitt G, Melzer D: The current landscape for direct-to-consumer genetic testing: legal, ethical, and policy issues. Annu Rev Genomics Hum Genet 2008;9:161–182.

43 Rantanen E, Hietala M, Kristoffersson U, Nippert I, Schmidtke J, Sequeiros J, Kaariainen H: What is ideal genetic counselling? A survey of current international guidelines. Eur J Hum Genet 2008;16:445–452.
44 Osterweil N: Full court press on hoop star Curry to get DNA testing. MedPage Today 2005. http://www.medpagetoday.com/Cardiology/Arrhythmias/1843 (accessed May 10, 2015).
45 Hesketh T, Xing ZW: Abnormal sex ratios in human populations: causes and consequences. Proc Natl Acad Sci U S A 2006;103:13271–13275.
46 Heffernan SM, Kilduff LP, Day SH, Pitsiladis YP, Williams AG: Genomics in rugby union: a review and future prospects. Eur J Sport Sci 2015;15:460–468.
47 Williams AG, Day SH, Lockey SJ, Heffernan SM, Erskine RM: Genomics as a practical tool in sport – have we reached the starting line? Cell Mol Exerc Physiol 2014;3:e6.

Alun G. Williams
MMU Sports Genomics Laboratory
Manchester Metropolitan University
Crewe Green Road, Crewe CW1 5DU (UK)
E-Mail a.g.williams@mmu.ac.uk

Author Index

Subject Index